MTEL Physics

11 Teacher Certification Exam

By: Sharon Wynne, M.S
Southern Connecticut State University

"And, while there's no reason yet to panic, I think it's only prudent that we make preparations to panic."

XAMonline, INC.

Boston

To obtain permission(s) to use the material from this work for any purpose including workshops or seminars, please submit a written request to:

XAMonline, Inc.
21 Orient Ave.
Melrose, MA 02176
Toll Free 1-800-509-4128
Email: info@xamonline.com
Web www.xamonline.com
Fax: 1-781-662-9268

Library of Congress Cataloging-in-Publication Data

Wynne, Sharon A.
     Physics 123-127: Teacher Certification / Sharon A. Wynne. -2nd ed.
     ISBN 978-1-58197-041-8
     1. Physics 123, 127.   2. Study Guides.   3. CSET
     4. Teachers' Certification & Licensure.   5. Careers

Disclaimer:
The opinions expressed in this publication are the sole works of XAMonline and were created independently from the National Education Association, Educational Testing Service, or any State Department of Education, National Evaluation Systems or other testing affiliates.

Between the time of publication and printing, state specific standards as well as testing formats and website information may change that is not included in part or in whole within this product. Sample test questions are developed by XAMonline and reflect similar content as on real tests; however, they are not former tests. XAMonline assembles content that aligns with state standards but makes no claims nor guarantees teacher candidates a passing score. Numerical scores are determined by testing companies such as NES or ETS and then are compared with individual state standards. A passing score varies from state to state.

Printed in the United States of America   œ-1

MTEL: Physics 11
ISBN: 978-1-58197-041-8

Table of Contents

Great Study and Testing Tips!

What to study in order to prepare for the subject assessments is the focus of this study guide but equally important is *how* you study.

You can increase your chances of truly mastering the information by taking some simple, but effective steps.

Study Tips:

1. <u>Some foods aid the learning process</u>. Foods such as milk, nuts, seeds, rice, and oats help your study efforts by releasing natural memory enhancers called CCKs (*cholecystokinin*) composed of *tryptopha*n, *choline*, and *phenylalanine*. All of these chemicals enhance the neurotransmitters associated with memory. Before studying, try a light, protein-rich meal of eggs, turkey, and fish. All of these foods release the memory enhancing chemicals. The better the connections, the more you comprehend.

Likewise, before you take a test, stick to a light snack of energy boosting and relaxing foods. A glass of milk, a piece of fruit, or some peanuts all release various memory-boosting chemicals and help you to relax and focus on the subject at hand.

2. <u>Learn to take great notes</u>. A by-product of our modern culture is that we have grown accustomed to getting our information in short doses (i.e. TV news sound bites or USA Today style newspaper articles.)

Consequently, we've subconsciously trained ourselves to assimilate information better in <u>neat little packages</u>. If your notes are scrawled all over the paper, it fragments the flow of the information. Strive for clarity. Newspapers use a standard format to achieve clarity. Your notes can be much clearer through use of proper formatting. A very effective format is called the <u>*"Cornell Method."*</u>

Take a sheet of loose-leaf lined notebook paper and draw a line all the way down the paper about 1-2" from the left-hand edge.

Draw another line across the width of the paper about 1-2" up from the bottom. Repeat this process on the reverse side of the page.

Look at the highly effective result. You have ample room for notes, a left hand margin for special emphasis items or inserting supplementary data from the textbook, a large area at the bottom for a brief summary, and a little rectangular space for just about anything you want.

3. **Get the concept then the details.** Too often we focus on the details and don't gather an understanding of the concept. However, if you simply memorize only dates, places, or names, you may well miss the whole point of the subject.

A key way to understand things is to put them in your own words. If you are working from a textbook, automatically summarize each paragraph in your mind. If you are outlining text, don't simply copy the author's words.

Rephrase them in your own words. You remember your own thoughts and words much better than someone else's, and subconsciously tend to associate the important details to the core concepts.

4. **Ask Why?** Pull apart written material paragraph by paragraph and don't forget the captions under the illustrations.

Example: If the heading is "Stream Erosion", flip it around to read "Why do streams erode?" Then answer the questions.

If you train your mind to think in a series of questions and answers, not only will you learn more, but it also helps to lessen the test anxiety because you are used to answering questions.

5. **Read for reinforcement and future needs.** Even if you only have 10 minutes, put your notes or a book in your hand. Your mind is similar to a computer; you have to input data in order to have it processed. *By reading, you are creating the neural connections for future retrieval.* The more times you read something, the more you reinforce the learning of ideas.

Even if you don't fully understand something on the first pass, *your mind stores much of the material for later recall.*

6. **Relax to learn so go into exile.** Our bodies respond to an inner clock called biorhythms. Burning the midnight oil works well for some people, but not everyone.

If possible, set aside a particular place to study that is free of distractions. Shut off the television, cell phone, and pager and exile your friends and family during your study period.

If you really are bothered by silence, try background music. Light classical music at a low volume has been shown to aid in concentration over other types. Music that evokes pleasant emotions without lyrics is highly suggested. Try just about anything by Mozart. It relaxes you.

7. <u>**Use arrows not highlighters.**</u> At best, it's difficult to read a page full of yellow, pink, blue, and green streaks. Try staring at a neon sign for a while and you'll soon see that the horde of colors obscure the message.

A quick note, a brief dash of color, an underline, and an arrow pointing to a particular passage is much clearer than a horde of highlighted words.

8. <u>**Budget your study time.**</u> Although you shouldn't ignore any of the material, *allocate your available study time in the same ratio that topics may appear on the test.*

Testing Tips:

1. Get smart, play dumb. Don't read anything into the question. Don't make an assumption that the test writer is looking for something else than what is asked. Stick to the question as written and don't read extra things into it.

2. Read the question and all the choices _twice_ before answering the question. You may miss something by not carefully reading, and then re-reading both the question and the answers.

If you really don't have a clue as to the right answer, leave it blank on the first time through. Go on to the other questions, as they may provide a clue as to how to answer the skipped questions.

If later on, you still can't answer the skipped ones . . . _Guess._ The only penalty for guessing is that you _might_ get it wrong. Only one thing is certain; if you don't put anything down, you will get it wrong!

3. Turn the question into a statement. Look at the way the questions are worded. The syntax of the question usually provides a clue. Does it seem more familiar as a statement rather than as a question? Does it sound strange?

By turning a question into a statement, you may be able to spot if an answer sounds right, and it may also trigger memories of material you have read.

4. Look for hidden clues. It's actually very difficult to compose multiple-foil (choice) questions without giving away part of the answer in the options presented. In most multiple-choice questions you can often readily eliminate one or two of the potential answers. This leaves you with only two real possibilities and automatically your odds go to Fifty-Fifty for very little work.

5. Trust your instincts. For every fact that you have read, you subconsciously retain something of that knowledge. On questions that you aren't really certain about, go with your basic instincts. **Your first impression on how to answer a question is usually correct.**

6. Mark your answers directly on the test booklet. Don't bother trying to fill in the optical scan sheet on the first pass through the test.

Just be very careful not to miss-mark your answers when you eventually transcribe them to the scan sheet.

7. Watch the clock! You have a set amount of time to answer the questions. Don't get bogged down trying to answer a single question at the expense of 10 questions you can more readily answer.

THIS PAGE BLANK

SUBAREA I. _____ **SCIENTIFIC INQUIRY**

COMPETENCY 1.0   **UNDERSTANDS THE HISTORICAL AND CONTEMPORARY CONTEXTS OF THE STUDY OF PHYSICS AND THE APPLICATIONS OF PHYSICS TO EVERYDAY LIFE**

Skill 1.1   **Explain the significance of key individuals and their theories in the history of physics**

Archimedes

Archimedes was a Greek mathematician, physicist, engineer, astronomer, and philosopher. He is credited with many inventions and discoveries some of which are still in use today such as the Archimedes screw. He designed the compound pulley, a system of pulleys used to lift heavy loads such as ships.

Although Archimedes did not invent the lever, he gave the first rigorous explanation of the principles involved which are the transmission of force through a fulcrum and moving the effort applied through a greater distance than the object to be moved. His Law of the Lever states that magnitudes are in equilibrium at distances reciprocally proportional to their weights.

He also laid down the laws of flotation and described Archimedes' principle which states that a body immersed in a fluid experiences a buoyant force equal to the weight of the displaced fluid.

Niels Bohr

Bohr was a Danish physicist who made fundamental contributions to understanding atomic structure and quantum mechanics. Bohr is widely considered one of the greatest physicists of the twentieth century. Bohr's model of the atom was the first to place electrons in discrete quantized orbits around the nucleus.

Bohr also helped determine that the chemical properties of an element are largely determined by the number of electrons in the outer orbits of the atom. The idea that an electron could drop from a higher-energy orbit to a lower one emitting a photon of discrete energy originated with Bohr and became the basis for future quantum theory.

He also contributed significantly to the Copenhagen interpretation of quantum mechanics. He received the Nobel Prize for Physics for this work in 1922.

Marie Curie

Curie was as a Polish-French physicist and chemist. She was a pioneer in radioactivity and the winner of two Nobel Prizes, one in Physics and the other in Chemistry. She was also the first woman to win the Nobel Prize.

Curie studied radioactive materials, particularly pitchblende, the ore from which uranium was extracted. The ore was more radioactive than the uranium extracted from it which led the Curies (Marie and her husband Pierre) to discover a substance far more radioactive then uranium. Over several years of laboratory work the Curies eventually isolated and identified two new radioactive chemical elements, polonium and radium. Curie refined the radium isolation process and continued intensive study of the nature of radioactivity.

Albert Einstein

Einstein was a German-born theoretical physicist who is widely considered one of the greatest physicists of all time. While best known for the theory of relativity, and specifically mass-energy equivalence, $E = mc^2$, he was awarded the 1921 Nobel Prize in Physics for his explanation of the photoelectric effect and "for his services to Theoretical Physics". In his paper on the photoelectric effect, Einstein extended Planck's hypothesis ($E = h\nu$) of discrete energy elements to his own hypothesis that electromagnetic energy is absorbed or emitted by matter in quanta and proposed a new law $E_{max} = h\nu - P$ to account for the photoelectric effect.

He was known for many scientific investigations including the special theory of relativity which stemmed from an attempt to reconcile the laws of mechanics with the laws of the electromagnetic field. His general theory of relativity considered all observers to be equivalent, not only those moving at a uniform speed. In general relativity, gravity is no longer a force, as it is in Newton's law of gravity, but is a consequence of the curvature of space-time.

Other areas of physics in which Einstein made significant contributions, achievements or breakthroughs include relativistic cosmology, capillary action, critical opalescence, classical problems of statistical mechanics and problems in which they were merged with quantum theory (leading to an explanation of the Brownian movement of molecules), atomic transition probabilities, the quantum theory of a monatomic gas, the concept of the photon, the theory of radiation (including stimulated emission), and the geometrization of physics.

Einstein's research efforts after developing the theory of general relativity consisted primarily of attempts to generalize his theory of gravitation in order to unify and simplify the fundamental laws of physics, particularly gravitation and electromagnetism, which he referred to as the Unified Field Theory.

Michael Faraday

Faraday was an English chemist and physicist who contributed significantly to the fields of electromagnetism and electrochemistry. He established that magnetism could affect rays of light and that the two phenomena were linked. It was largely due to his efforts that electricity became viable for use in technology. The unit for capacitance, the farad, is named after him as is the Faraday constant, the charge on a mole of electrons (about 96,485 coulombs). Faraday's law of induction states that a magnetic field changing in time creates a proportional electromotive force.

J. Robert Oppenheimer

Oppenheimer was an American physicist best known for his role as the scientific director of the Manhattan Project, the effort to develop the first nuclear weapons. Sometimes called "the father of the atomic bomb", Oppenheimer later lamented the use of atomic weapons. He became a chief advisor to the United States Atomic Energy Commission and lobbied for international control of atomic energy. Oppenheimer was one of the founders of the American school of theoretical physics at the University of California, Berkeley. He did important research in theoretical astrophysics, nuclear physics, spectroscopy, and quantum field theory.

Sir Isaac Newton

Newton was an English physicist, mathematician, astronomer, alchemist, and natural philosopher in the late 17th and early 18th centuries. He described universal gravitation and the three laws of motion laying the groundwork for classical mechanics. He was the first to show that the motion of objects on earth and in space is governed by the same set of mechanical laws. These laws became central to the scientific revolution that took place during this period of history. Newton's three laws of motion are:

> I. Every object in a state of uniform motion tends to remain in that state of motion unless an external force is applied to it.
> II. The relationship between an object's mass m, its acceleration a, and the applied force F is $F = ma$.
> III. For every action there is an equal and opposite reaction.

In mechanics, Newton developed the basic principles of conservation of momentum. In optics, he invented the reflecting telescope and discovered that the spectrum of colors seen when white light passes through a prism is inherent in the white light and not added by the prism as previous scientists had claimed. Newton notably argued that light is composed of particles. He also formulated an experimental law of cooling, studied the speed of sound, and proposed a theory of the origin of stars.

Erwin Schrödinger

Schrödinger was a Nobel Prize winning Austrian physicist who is best remembered for his contributions to quantum mechanics. Chief among his findings was the Schrödinger equation which describes the space and time-dependence of quantum systems. This equation predicts behavior of microscopic particles in much the same way that Newton's second law predicts the behavior of macroscopic matter. Much of solid state physics was built upon the Schrödinger (wave) equation. Schrödinger also devised the "cat in a box" thought experiment to illustrate a paradox of quantum mechanics.

Enrico Fermi

Fermi was an extremely gifted Italian physicist and Nobel Prize winner. His work focused on radioactivity, quantum theory, and statistical mechanics and he helped develop the first nuclear reactor. He also participated in the Manhattan Project after fleeing to America prior to the Second World War. Fermi's Golden Rule is an important equation in quantum mechanics and is used to calculate the transition rates of quantum systems. Much of Fermi's work in quantum mechanics (Fermi holes, Fermi levels, Fermi-Dirac statistics) ultimately led to our modern understanding of semiconductors.

James Clerk Maxwell

Maxwell was a Scottish theoretical physicist and mathematician whose work was especially important to our understanding of electromagnetism. His famous Maxwell's equations were the first to unify and simply express the laws of electricity and magnetism. Maxwell was also important in the development of the kinetic theory; the Maxwell-Boltzman distribution is a probability distribution used in statistical mechanics and to predict the behavior of gases. His work paved the way for much of the progress in physics during the 20th century (special relativity, quantum mechanics, etc) and ultimately led to all modern electrical and communications technology.

Daniel Bernoulli

Bernoulli was a Swiss mathematician who developed many equations important in physics. His work was especially focused on the behavior of fluids and the Bernoulli equation governs all steady, inviscid, incompressible flow. The resultant Bernoulli principle is especially useful in aerodynamics. The Euler-Bernoulli beam theory, which allows calculation of the loading characteristics of a beam, was of key importance during the Industrial Revolution and remains an important equation for engineers today. Bernoulli was also the first scientist to suggest the principles of the kinetic theory of gases.

Skill 1.2 **Explain the societal implications of developments in physics (e.g., nuclear technology, solid state technology)**

Nearly all advances in science and technology have some effect on society. However, some are especially important in changing the way we think about the natural world or how we lead our day to day lives. Several examples drawn from physics are briefly explored below.

Heliocentric theory

While we largely take for granted that the planets in our solar system orbit around the sun, during the 16th and 17th centuries it was a controversial notion in the western world. Most religious authorities, at least publicly, condemned the writings of Galileo and Copernicus which postulated a heliocentric model of the solar system. Both scientists and theologians argued over how new scientific evidence could be brought into agreement with scripture. This provides a famous historical example of how organized religion can interact with, and at times impede, scientific progress. Today, we still see this playing out in debates over such theories as the Big Bang and evolution.

Nuclear physics and the atomic bomb

Advances in nuclear physics in the first half of the 20th century led to the realization that tremendous amounts of power could be generated for fuel or weaponry through atomic fission and fusion. During World War II, the United State and its allies (through the Manhattan Project) were the first to succeed in the creation of atomic bombs which were subsequently dropped on Hiroshima and Nagasaki in 1945. The long term and exceedingly devastating effect of these weapons cannot be overstated. Nuclear proliferation (especially by the US and the USSR) in the following 50 years contributed to the Cold War and created an imminent threat of vast worldwide destruction. While the Cold War has since ended, the continued testing and possession of nuclear weaponry by an ever growing number of countries remains an important socio-political problem.

Nuclear power is currently use but remains controversial. On one hand, the extreme short and long term dangers of nuclear accidents have been demonstrated by incidents such as those at Three Mile Island and Chernobyl. On the other hand, society is highly in need of a cleaner, more sustainable source of energy and nuclear power might be able to fill this role.

Semiconductors, the transistor, and digital devices

The discovery that the electrical conductive properties of semiconductors can be either permanently or temporarily manipulated has been an important one. Most importantly, it led to the creation of solid state electrical devices such as transistors. In electrical circuits, these elements replaced the previously used vacuum tubes and mechanical relays which were highly susceptible to wear and physical stressors. Transistors, however, are small, fast, accurate, reliable, and can be produced quickly and cheaply. Thus, transistors have paved for the way for the vast array of affordable electrical devices. They have been particularly important in advancing computers and digital technology. We are well aware of how digital devices and information technology, made possible in part by transistors, touch every part of our modern lives.

MEMS and nanotechnology

Many people speculate that nanotechnology is poised to make significant changes in our lives. Microelectromechanical Systems (MEMS), or micromachines are made possible by modern materials and technologies (such as lasers and micro-etching) that allow manufacture and manipulation of devices on the micrometer scale. Developments in these technologies are already being used in the automotive, biomedical, chemical, cosmetic, telecommunications, and manufacturing industries.

COMPETENCY 2.0   UNDERSTAND THE NATURE OF SCIENTIFIC INQUIRY, SCIENTIFIC PROCESSES, AND THE ROLE OF OBSERVATION AND EXPERIMENTATION IN SCIENCE

Skill 2.1   Outline the processes by which new scientific knowledge and hypotheses are generated

The scientific method is a logical set of steps that a scientist goes through to solve a problem. There are as many different scientific methods as there are scientists experimenting. However, there seems to be some pattern to their work. The scientific method is the process by which data is collected, interpreted and validated. While an inquiry may start at any point in this method and may not involve all of the steps here is the general pattern.

Formulating problems
Although many discoveries happen by chance, the standard thought process of a scientist begins with forming a question to research. The more limited and clearly defined the question, the easier it is to set up an experiment to answer it. Scientific questions result from observations of events in nature or events observed in the laboratory. An **observation** is not just a look at what happens. It also includes measurements and careful records of the event. Records could include photos, drawings, or written descriptions. The observations and data collection lead to a question. In physics, observations almost always deal with the behavior of matter. Having arrived at a question, a scientist usually researches the scientific literature to see what is known about the question. Maybe the question has already been answered. The scientist then may want to test the answer found in the literature. Or, maybe the research will lead to a new question.

Sometimes the same observations are made over and over again and are always the same. For example, you can observe that daylight lasts longer in summer than in winter. This observation never varies. Such observations are called **laws** of nature. One of the most important scientific laws was discovered in the late 1700s. Chemists observed that no mass was ever lost or gained in chemical reactions. This law became known as the law of conservation of mass. Explaining this law was a major topic of scientific research in the early 19th century.

Forming a hypothesis

Once the question is formulated, take an educated guess about the answer to the problem or question. This 'best guess' is your hypothesis. A **hypothesis is a statement of a possible answer to the question**. It is a tentative explanation for a set of facts and can be tested by experiments. Although hypotheses are usually based on observations, they may also be based on a sudden idea or intuition.

Experiment

An experiment tests the hypothesis to determine whether it may be a correct answer to the question or a solution to the problem. Some experiments may test the effect of one thing on another under controlled conditions. Such experiments have two variables. The experimenter controls one variable, called the **independent variable**. The other variable, the **dependent variable**, is the change caused by changing the independent variable. For example, suppose a researcher wanted to test the effect of vitamin A on the ability of rats to see in dim light. The independent variable would be the dose of Vitamin A added to the rats' diet. The dependent variable would be the intensity of light that causes the rats to react. All other factors, such as time, temperature, age, water given to the rats, the other nutrients given to the rats, and similar factors, are held constant. Scientists sometimes do short experiments "just to see what happens". Often, these are not formal experiments. Rather they are ways of making additional observations about the behavior of matter. A good test will try to manipulate as few variables as possible so as to see which variable is responsible for the result. This requires a second example of a **control**. A control is an extra setup in which all the conditions are the same except for the variable being tested.

In most experiments, scientists collect quantitative data, which is data that can be measured with instruments. They also collect qualitative data, descriptive information from observations other than measurements. Interpreting data and analyzing observations are important. If data is not organized in a logical manner, wrong conclusions can be drawn. Also, other scientists may not be able to follow your work or repeat your results.

Conclusion

Finally, a scientist must draw conclusions from the experiment. A conclusion must address the hypothesis on which the experiment was based. The conclusion states whether or not the data supports the hypothesis. If it does not, the conclusion should state what the experiment *did* show. If the hypothesis is not supported, the scientist uses the observations from the experiment to make a new or revised hypothesis. Then, new experiments are planned.

Theory

When a hypothesis survives many experimental tests to determine its validity, the hypothesis may evolve into a **theory**. A theory explains a body of facts and laws that are based on the facts. A theory also reliably predicts the outcome of related events in nature. For example, the law of conservation of matter and many other experimental observations led to a theory proposed early in the 19th century. This theory explained the conservation law by proposing that all matter is made up of atoms which are never created or destroyed in chemical reactions, only rearranged. This atomic theory also successfully predicted the behavior of matter in chemical reactions that had not been studied at the time. As a result, the atomic theory has stood for 200 years with only small modifications.

A theory also serves as a scientific **model**. A model can be a physical model made of wood or plastic, a computer program that simulates events in nature, or simply a mental picture of an idea. A model illustrates a theory and explains nature. For instance, in your science class you may develop a mental (and maybe a physical) model of the atom and its behavior. Outside of science, the word theory is often used to describe someone's unproven notion about something. In science, theory means much more. It is a thoroughly tested explanation of things and events observed in nature. A theory can never be proven true, but it can be proven untrue. All it takes to prove a theory untrue is to show an exception to the theory. The test of the hypothesis may be observations of phenomena or a model may be built to examine its behavior under certain circumstances.

Steps of a Scientific Method

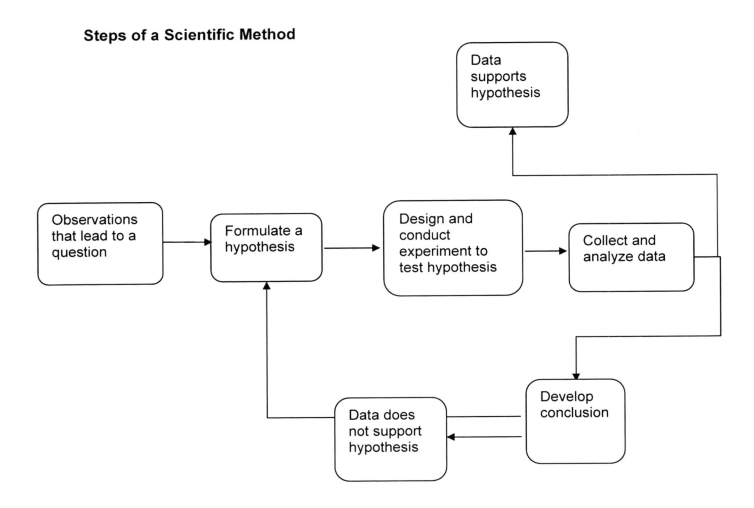

Skill 2.2 **Give examples of ethical issues related to scientific processes (e.g., accurately reporting experimental results)**

Ethical behavior is critical at all points of scientific experimentation, interpretation, and communication. Both other scientists and the public in general need to be confident in the validity of scientific results and the fact that science is being performed in the service of public good.

The first aspect of this is that **science is actually performed responsibly**. It is always important that careful documents be kept of all experimental conditions and outcomes. In both commercial and academic research environments, scientists keep careful records in laboratory notebooks. These serve as primary documents bearing witness to new discoveries and developments. Additionally, it is of utmost importance that, when used, animal subjects are humanely treated and human subjects are fully aware of all relevant facts (informed consent). Most research and funding organizations have policies in place and review boards to monitor the use of animal and human subjects.

Secondly, the **scientific findings must be accurately presented**. This means that all relevant data is presented to clearly reflect experimental findings. Occasionally, a scientist will report only that data which agrees with the hypothesis he is putting forth. Alternatively, the data may be improperly statistically manipulated. These and other misrepresentations of experimental findings seriously impede the progress of science because they may provide misinformation about how the natural world operates.

Another aspect of the requirement to accurately present research is the need for the **credit for work to be properly assigned**. Whether in the development of technology or in pure research, there are both financial and scientific issues at stake. In academic and pure research settings findings are typically published in peer reviewed journals and student theses. Occasionally, cases of plagiarism or even the theft of experimental work occur. Plagiarism means that the work has been presented as original when, in fact, it incorporates the work of others. Both these occurrences are highly undesirable and may results in disciplinary action and/or loss of reputation within the scientific community. It should be noted that, when preparing documents, it is acceptable to include information and even direction quotations from other sources as long as they are always *properly cited*.

One final area in which ethics is extremely important is in **deciding in what direction scientific research should proceed**. Both scientists and the general public must determine if there are any possible investigations that *should* not actually be performed. Currently, this is especially important in medical research where appropriateness of research into human cloning, embryonic stems cells, and other topics are heavily debated.

Skill 2.3   Evaluating the appropriateness of a specified experimental design to test a given physics hypothesis

Disturbance by observation

Experiments designed to test physics hypotheses are plagued by a number of difficulties that can be partially, but never totally, alleviated. For example, as has been pointed out by many, an experiment does not involve a scientist making an observation of a phenomenon; rather, it involves a scientist making an observation of a phenomenon being observed by a scientist. That is to say, the act of observation has an effect on the system being observed and this must be recognized in analyzing and reporting results. This effect can be profound in microscopic experiments, especially at the level of subatomic particles where using light to observe a particle actually disturbs the particle substantially. A macroscopic example is the measurement of voltage or current in an electrical circuit. Although most voltmeters and ammeters do not present a large load to the circuit, they do load the circuit to some degree affecting the measurements. Thus, it is critical to note the inherent limitations present in an experiment when evaluating its ability to properly test some hypothesis.

Isolation of parameters

Along similar lines to the above-mentioned difficulty, it is not possible to completely isolate a particular parameter for experimental testing. Thus, for example, although theory can treat a mass as the sole entity in the universe, in reality there are virtually innumerable other masses distributed throughout the universe that have an effect (by gravitation, for example) on the mass of interest. An appropriate experiment is thus designed to minimize outside and unwanted influences to the extent possible. One should evaluate how well the experiment tests only those parameters that are of concern to the hypothesis and how well the experiment reduces, or holds invariable, other parameters that could otherwise affect the results.

Measurement precision

Another concern regarding experimental design is the ability of the equipment to measure the desired parameters with sufficient accuracy and precision. If the parameter that is being measured is beyond the range of the equipment by way of being either too large, too small, or having variations that are too fine, the experiment can yield no useful information. The measurement equipment may be more or less technical; it may simply involve the use of the human eye or it may involve sophisticated equipment such as radiation or thermal energy detectors, pressure sensors or other devices.

Financial burden

Financial concerns are no less important than physical or scientific concerns. An experiment is only as useful as the possibility of its implementation and the cost of an experiment is a critical consideration in its appropriate design. High-energy physics, for example, is hampered by this difficulty as the costs of constructing particle accelerators and other associated equipment, not to mention the costs of maintenance and personnel, can be staggeringly high. Thus, an ideal experimental design is one that can be implemented at an affordable level of expense that allows for repeated use and proper maintenance.

Skill 2.4   **Recognizing the role of communication among scientists and between scientists and the public in promoting scientific progress**

There are many important reasons for scientists to report their detailed findings to the scientific community. The first and most important reason is to **ensure the correctness of scientific results and to advance our larger understanding of the natural world.** When new scientific findings are reported at conferences or peer reviewed for publication in technical journals, they are rigorously evaluated. This is to ensure that the experiments were performed with proper controls and that the results are repeatable. In controversial situations, experiments may even be fully repeated by other scientists to ensure that the same results are obtained. Once it is established that new findings are sound, the scientists can work together to determine how these results agree or disagree with previous experiments. They may also decide together what additional investigations would be useful to the field. **When open-minded scientific discourse is supported, opinions and information can be exchanged and theories can be refined.** Scientists will weigh the experimental evidence to determine what hypothesis has the most support. Given time, new theories may emerge and scientific knowledge can both broaden and deepen.

Science has increasingly become a collaborative and interdisciplinary effort. Engineers, clinicians, and scientists may not be familiar with advances in far reaching fields unless new developments are announced to a broad community of scientists. When this happens, it opens the doors for an exchange of information that can be enormously helpful in both investigating and solving problems in science and technology. Consider, for instance, an exchange of information between physicists and biologists. The physicists may suggest the use of high powered microscopes and lasers for study of sub-cellular structures. Likewise, the biologist's knowledge of transport proteins could inform the physicist's design of new nano-machines.

It is also important that the public in general be kept informed of scientific progress. This is of key importance in any democratic society because all people together must decide the direction of the government. **New scientific discoveries can have huge implications for medical, social, and environmental issues which should be of interest to all voters.** Additionally, when people are made familiar with the advances made in science, they may be more willing to see the need for public funding which supports continued scientific progress.

COMPETENCY 3.0   UNDERSTAND THE PROCESSES OF
                 GATHERING, ORGANIZING, REPORTING, AND
                 ANALYZING SCIENTIFIC DATA IN THE CONTEXT OF
                 PHYSICS INVESTIGATIONS

Skill 3.1   Evaluating the appropriateness of a given method or
             procedure for collecting data

Resistance to human error
A number of factors influence the appropriateness of data collection methods
with regard to the purpose for which they are being employed. The likelihood of
human error must be screened carefully, as mistakes in setting up equipment,
transcription errors in recording data or disturbances of the setup during data
collection can all lead to skewed results. In some situations, it may be sufficient
for the scientist to simply take measurements and record them by hand, but, at
other times, a computer or data acquisition system may be required. In many
electromagnetics applications, for example, the data rates or frequencies of
operation are often so high as to make automated data collection a necessity.

Sufficient sampling
The data collection method must be able sample data at a sufficiently high rate to
adequately characterize the process or phenomenon being studied. In the case
of transient wave phenomena, such as transient electromagnetic or acoustic
signals, the data collection rate must be at least twice as high as the highest
frequency component of the signal. This so-called Nyquist rate is the minimum
threshold for preventing aliasing in the collected data. The minimum
requirements for data sampling may be temporal, as with many acoustic and
electromagnetic signals, or they may be spatial, as with remote sensing, for
example.

Sufficient data storage
Along similar lines, sufficient storage space for data must be available. Often, this
involves computer storage in the form of a hard drive, or, perhaps, an optical
medium. Data acquisition systems operating at high rates can produce
tremendous amounts of data in short periods of time. Financial and even space
considerations apply when a large number of CDs, DVDs or other media are
required for storage.

Data precision

The precision of the data collection approach must also be evaluated. Both the sensing or measurement device, as well as the digitization equipment (if computer-based data acquisition is being used), must be of sufficient precision to measure the parameters of the system of interest to accurately record variations in those parameters. If, for example, a transducer is too coarse in its measurement abilities or an analog-to-digital converter does not have a sufficient number of bits to accurately digitize the signal, then information is lost. Similarly, the data that is collected should not be assigned more precision than the equipment allows. If the digitization equipment is more precise than the measurement or transduction equipment, for example, then the data collected may have more numerical precision than the transducer could have possibly afforded.

Skill 3.2 Selecting an appropriate and effective graphic representation for organizing and reporting experimental data

In scientific investigations, it is often necessary to gather and analyze large data sets. We may need to manage data taken over long periods of time and under various different conditions. Appropriately organizing this data is necessary to identify trends and present the information to others. The uses and advantages of various graphic representations are discussed below.

Tables

Tables are excellent for organizing data as it is being recorded and for storing data that needs to be analyzed. In fact, almost all experimental data is initially organized into a table, such as in a lab notebook. Often, it is then entered into tables within spreadsheets for further processing. Tables can be used for presenting data to others if the data set is fairly small or has been summarized (for example, presenting average values). However, for larger data sets, tabular presentation may be overwhelming. Finally, tables are not particularly useful for recognizing trends in data or for making them apparent to others.

Charts and Graphs

Charts and graphs are the best way to demonstrate trends or differences between groups. They are also useful for summarizing data and presenting it. In most types of graphs, it is also simple to indicate uncertainty of experimental data using error bars. Many types of charts and graphs are available to meet different needs. Three of the most common are scatterplots, bar charts, and pie charts. An example of each is shown.

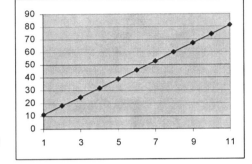

Scatterplots are typically shown on a Cartesian plane and are useful for demonstrating the relationship between two variables. A line chart (shown) is a special type of two-dimensional scatter plot in which the data points are connected with a line to make a trend more apparent.

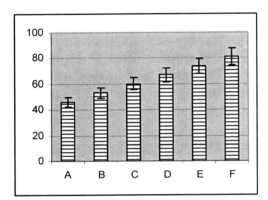

Bar charts can sometimes fill the same role as scatterplots but are better suited to show values across different categories or different experimental conditions (especially where those conditions are described qualitatively rather than quantified). Note the use of error bars in this example.

Finally, a pie chart is best used to present relative magnitudes or frequencies of several different conditions or events. They are most commonly used to show how various categories contribute to a whole.

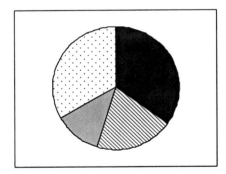

Diagrams

Diagrams are not typically used to present the specifics of data. However, they are very good for demonstrating phenomena qualitatively. Diagrams make it

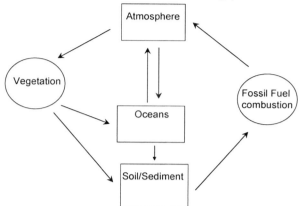

easy to visualize the connections and relationships between various elements. They may also be used to demonstrate temporal relationships. For example, diagrams can be used to illustrate the operation of an internal combustion engine or the complex biochemical pathways of an enzyme's action. The diagram shown is a simplified version of the carbon cycle.

Skill 3.3   Applying procedures and criteria for formally reporting experimental procedures and data to the scientific community

Once scientists have rigorously performed a set of experiments, they may believe that they have information of significant value that should be shared within the scientific community. **New findings are often presented at scientific conferences and ultimately published in technical journals** (some well known examples include *Nature*, *Science*, and the *Journal of the American Medical Association*). Scientists prepare manuscripts detailing the conditions of their experiments and the results they obtained. They will typically also include their interpretation of those results and their impact on current theories in the field. These manuscripts are not wholly unlike lab reports, though they are considerably more polished, of course. Manuscripts are then submitted to appropriate technical journals.

All reputable scientific journals use peer review to assess the quality of research submitted for publication. Peer review is the process by which scientific results produced by one person or group are subjected to the analysis of other experts in the field. It is important that they be objective in their evaluations. Peer review is typically done anonymously so that the identity of the reviewer remains unknown by the scientists submitting work for review. The goal of peer review is to "weed out" any science not performed by appropriate standards. Reviewers will determine whether proper controls were in place, enough replicates were performed, and that the experiments clearly address the presented hypothesis. The reviewer will scrutinize the interpretations and how they fit into what is already known in the field. Often reviewers will suggest that additional experiments be done to further corroborate presented conclusions. If the reviewers are satisfied with the quality of the work, it will be published and made available to the entire scientific community.

Sometimes, there is significant opposition to new ideas, especially if they conflict with long-held ideas in a scientific field. However, with enough correct experimental support, scientific theories will in time be adjusted to reflect the newer findings. Thus science is an ongoing cycle: ideas are constantly being refined and modified to reflect new evidence and, ultimately, provide us with a more correct model of the world around us. There are many scientific theories that have experienced multiple revisions and expansions. Some well-known examples include the atomic theory, which was notable refined by Dalton, Thomson, Rutherford, and Bohr and the theory of natural selection, which was originally formulated by Darwin but has been enriched by discoveries in molecular biology.

Skill 3.4   Explain relationships between factors (e.g., linear, direct, inverse, direct squared, inverse squared) as indicated by experimental data

In many experimental investigations we attempt to understand the relationship between two variables. Typically, one variable (independent, usually represented x) is altered while the other is measured (dependent, usually represented y). Certain relationships between these variables are commonly observed. Often, the relationships are most clearly observable when graphed.

Linear
In linear relationships, the independent variable is *directly* proportional to the dependent variable. This means that, as the absolute value of x is increased, the absolute value of y will also increase. This is stated mathematically as:

$$y \propto x$$

This relationship can also be represented using the familiar equation for a straight line (where m is slope and b is the intercept):

$$y = mx + b$$

Graphs of this type of relationship are shown below. On the left, the slope of the line (m) is positive and on the right it is negative:

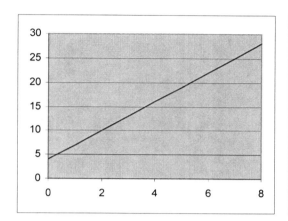 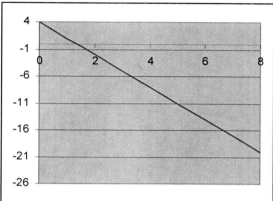

Inverse

Unlike direct proportionality, inverse proportionality signifies that the dependent variable is a function of the inverse of the independent variable:

$$y \propto \frac{1}{x}$$

A graph of this relationship is hyperbolic, as shown below:

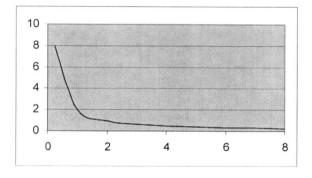

Squared and Inverse Squared

As the name suggests, in squared relationships the dependent variable is proportional to the square of the independent variable:

$$y \propto x^2$$

Similarly to the inverse proportionality described above, the dependent variable may also be proportional to the inverse square of the independent variable:

$$y \propto \frac{1}{x^2}$$

Graphs of the squared (left) and inversely squared (right) proportionality are shown below. Notice that the graphs are non-linear.

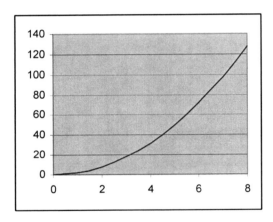 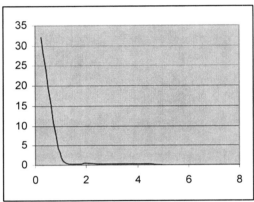

Logarithmic and Exponential

The dependent variable may also be proportional to a logarithmic or exponential function of the independent variable.

$$y \propto \log_a x \qquad y \propto a^x$$

Graphs of these relationships are shown below (logarithmic relationship on the left and exponential on the right). Make special note of the scale of the dependent variable.

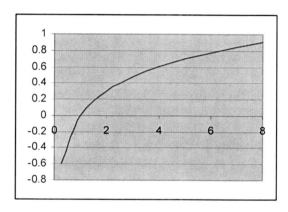
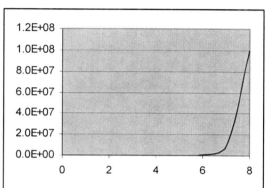

COMPETENCY 4.0   UNDERSTAND PRINCIPLES AND PROCEDURES OF
                  MEASUREMENT AND THE USE OF MATHEMATICS IN
                  PHYSICS (E.G., VECTOR ANALYSIS, CALCULUS)

Skill 4.1   Explain the importance of SI units

SI is an abbreviation of the French *Système International d'Unités* or the
International System of Units. It is the most widely used system of units in the
world and is the system used in science. The use of many SI units in the United
States is increasing outside of science and technology.

SI units are commonly accepted among most scientists for collecting and
reporting measurements. In a few cases there are some exceptions with regard
to SI unit usage. For instance, in the heating and air conditioning industries, the
British thermal unit (BTU) is used in place of the Joule, the SI unit for energy. A
system of units is always based on certain conventions, such as the precise
definition of a kilogram, a meter or a second, and many other units, such as the
Newton (unit of force) or the Joule (unit of energy) are derived from these
definitions. No particular system of units is inherently better than another
although some may be more convenient to use. SI is often a preferred system
due to its ubiquity among scientists, its largely base-10 systematization and its
popular use by most countries around the world. Some scientists still use the
centimeter-gram-second (CGS) system of units due to its usefulness in
simplifying some equations and in reducing some physical constants to unity.

There are two types of SI units: **base units** and **derived units**. The base units
are:

Quantity	Unit name	Symbol
Length	meter	m
Mass	kilogram	kg
Amount of substance	mole	mol
Time	second	s
Temperature	kelvin	K
Electric current	ampere	A
Luminous intensity	candela	cd

The name "kilogram" occurs for the SI base unit of mass for historical reasons. Derived units are formed from the kilogram, but appropriate decimal prefixes are attached to the word "gram." Derived units measure a quantity that may be **expressed in terms of other units**. Some derived units important for physics are:

Derived quantity	Unit name	Expression in terms of other units	Symbol
Area	square meter	m^2	
Volume	cubic meter	m^3	
	liter	$dm^3 = 10^{-3}\ m^3$	L or l
Mass	unified atomic mass unit	$(6.022 \times 10^{23})^{-1}\ g$	u or Da
Time	minute	60 s	min
	hour	60 min = 3600 s	h
	day	24 h = 86400 s	d
Speed	meter per second	m/s	
Acceleration	meter per second squared	m/s^2	
Temperature*	degree Celsius	K	°C
Mass density	gram per liter	$g/L = 1\ kg/m^3$	
Force	newton	$m \cdot kg/s^2$	N
Pressure	pascal	$N/m^2 = kg/(m \cdot s^2)$	Pa
	standard atmosphere§	101325 Pa	atm
Energy, Work, Heat	joule	$N \cdot m = m^3 \cdot Pa = m^2 \cdot kg/s^2$	J
	nutritional calorie§	4184 J	Cal
Heat (molar)	joule per mole	J/mol	
Heat capacity, entropy	joule per kelvin	J/K	
Heat capacity (molar), entropy (molar)	joule per mole kelvin	J/(mol·K)	
Specific heat	joule per kilogram kelvin	J/(kg·K)	
Power	watt	J/s	W
Electric charge	coulomb	s·A	C
Electric potential, electromotive force	volt	W/A	V
Viscosity	pascal second	Pa·s	
Surface tension	newton per meter	N/m	

*Temperature differences in Kelvin are the same as those differences in degrees Celsius. To obtain degrees Celsius from Kelvin, subtract 273.15. Differentiate *m* and meters (m) by context.
§These are commonly used non-SI units.

Decimal multiples of SI units are formed by attaching a **prefix** directly before the unit and a symbol prefix directly before the unit symbol. SI prefixes range from 10^{-24} to 10^{24}. Common prefixes you are likely to encounter in physics are shown below:

Factor	Prefix	Symbol	Factor	Prefix	Symbol
10^9	giga—	G	10^{-1}	deci—	d
10^6	mega—	M	10^{-2}	centi—	c
10^3	kilo—	k	10^{-3}	milli—	m
10^2	hecto—	h	10^{-6}	micro—	μ
10^1	deca—	da	10^{-9}	nano—	n
			10^{-12}	pico—	p

Example: 0.0000004355 meters is 4.355×10^{-7} m or 435.5×10^{-9} m. This length is also 435.5 nm or 435.5 nanometers.

Example: Find a unit to express the volume of a cubic crystal that is 0.2 mm on each side so that the number before the unit is between 1 and 1000.

Solution: Volume is length X width X height, so this volume is $(0.0002 \text{ m})^3$ or $8 \times 10^{-12} \text{ m}^3$. Conversions of volumes and areas using powers of units of length must take the power into account. Therefore:

$$1 \text{ m}^3 = 10^3 \text{ dm}^3 = 10^6 \text{ cm}^3 = 10^9 \text{ mm}^3 = 10^{18} \text{ } \mu\text{m}^3,$$

The length 0.0002 m is $2 \times 10^2 \text{ } \mu\text{m}$, so the volume is also $8 \times 10^6 \text{ } \mu\text{m}^3$. This volume could also be expressed as $8 \times 10^{-3} \text{ mm}^3$. None of these numbers, however, is between 1 and 1000.

Expressing volume in liters is helpful in cases like these. There is no power on the unit of liters, therefore:

$$1 \text{ L} = 10^3 \text{ mL} = 10^6 \text{ } \mu\text{L} = 10^9 \text{ nL}.$$

Converting cubic meters to liters gives

$8 \times 10^{-12} \text{ m}^3 \times \dfrac{10^3 \text{ L}}{1 \text{ m}^3} = 8 \times 10^{-9} \text{ L}$. The crystal's volume is 8 nanoliters (8 nL).

Example: Determine the ideal gas constant, R, in L•atm/(mol•K) from its SI value of 8.3144 J/(mol•K).

Solution: One joule is identical to one m^3•Pa (see the table on the previous page).

$$8.3144 \, \frac{\text{m}^3 \bullet \text{Pa}}{\text{mol} \bullet \text{K}} \times \frac{1000 \text{ L}}{1 \text{ m}^3} \times \frac{1 \text{ atm}}{101325 \text{ Pa}} = 0.082057 \, \frac{\text{L} \bullet \text{atm}}{\text{mol} \bullet \text{K}}$$

The **order of magnitude** is a familiar concept in scientific estimation and comparison. It refers to a category of scale or size of an amount, where each category contains values of a fixed ratio to the categories before or after. The most common ratio is 10. Orders of magnitude are typically used to make estimations of a number. For example, if two numbers differ by one order of magnitude, one number is 10 times larger than the other. If they differ by two orders of magnitude the difference is 100 times larger or smaller, and so on. It follows that two numbers have the same order of magnitude if they differ by less than 10 times the size.

To estimate the order of magnitude of a physical quantity, you round the its value to the nearest power of 10. For example, in estimating the human population of the earth, you may not know if it is 5 billion or 12 billion, but a reasonable order of magnitude estimate is 10 billion. Similarly, you may know that Saturn is much larger than Earth and can estimate that it has approximately 100 times more mass, or that its mass is 2 orders of magnitude larger. The actual number is 95 times the mass of earth. Below are the dimensions of some familiar objects expressed in orders of magnitude.

Physical Item	Size	Order of Magnitude (meters)
Diameter of a hydrogen atom	100 picometers	10^{-10}
Size of a bacteria	1 micrometer	10^{-6}
Size of a raindrop	1 millimeter	10^{-3}
Width of a human finger	1 centimeter	10^{-2}
Height of Washington Monument	100 meters	10^{2}
Height of Mount Everest	10 kilometers	10^{4}
Diameter of Earth	10 million meters	10^{7}
One light year	1 light year	10^{16}

Skill 4.2   Give examples of typical measuring devices used in physics

See Skill 6.1 **for a description of common laboratory instruments.**

Skill 4.3   Proper methods of measurement for given situations

The approach to taking measurements for a particular experiment is determined by the particular phenomenon being measured, the context of measurement and the financial limitations imposed upon the experiment. Obviously, thermometers cannot be used for measuring distances, and the right equipment must be used for the appropriate measurement. Nevertheless, more subtle considerations abound, including the required accuracy and precision of measurements. If relatively coarse measurements are sufficient, there is often no need for sophisticated, expensive equipment. For example, if a rough measurement of weight is required, a mechanical balance scale may be all that is required. Measurements of higher precision using an advanced digital scale may be gratuitous, especially in light of other inexact measurements that may be taken during the experiment.

The particular method used for measurement is determined in large part by the theory or hypothesis being tested. Measurements in the realm of particle physics, where hypotheses are largely based on a synthesis of quantum mechanics and special relativity, do not allow for the use of approaches to measurement that assume classical mechanics and electrodynamics. Instead, the particular method used must be based on more firmly established principles and concepts from the theory. Additionally, certain types of measurements must be approached indirectly. That is to say, empirical information about some parameters of a system may actually require measurement of different parameters from which measurements the desired parameter can be calculated. Such indirect measurements may be required for phenomena that are newly discovered, that are microscopic in scale or that require extremely high-precision results.

Skill 4.4   Analyzing uncertainty in measurements

Scientific data can never be error-free. We can, however, gain useful information from our data by understanding what the sources of error are, how large they are and how they affect our results. Some errors are intrinsic to the measuring instrument, others are operator errors. Errors may be random (in any direction) or systematic (biasing the data in a particular way). In any measurement that is made, data must be quoted along with an estimate of the error in it.

Accuracy is a measure of how close to "correct" a measuring device or technique is. **Precision** is a measure of how similar repeated measurements from a given device or technique are. While the best devices or techniques will yield measurements that are both accurate and precise, it is possible to be accurate without being precise or to be precise without being accurate.

The classic analogy to demonstrate accuracy and precision is that of a bulls-eye. Accuracy alone is shown in the left example: the shots are all close to the center of the bulls-eye (the correct value). Precision alone is shown in the middle example: the shots are tightly clustered together. Both accuracy and precision are shown in the example on the right: the shots are tightly clustered near the center of the bulls-eye.

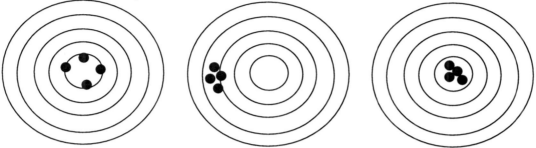

The accuracy of a technique or device can be determined in a straightforward manner. It is done by measuring a known quantity (a standard) and determining how close the value provided by the measuring device is.

To determine precision, we must make multiple measurements of the same sample. The precision of an instrument is typically given in terms of its standard error or standard deviation. Precision is typically divided into reproducibility and repeatability. These concepts are subtly different and are defined as follows:

Repeatability: Variation observed in measurements made over a short period of time while trying to keep all conditions the same (including using the same instrument, the same environmental conditions, and the same operator)

Reproducibility: Variation observed in measurements taken over a long time period in a variety of different settings (different places and environments, using different instruments and operators)

Both repeatability and reproducibility can be estimated by taking multiple measurements under the conditions specified above. Using the obtained values, standard deviation can be calculated using the formula:

$$\sigma = \sqrt{\frac{1}{N}\sum_{i=1}^{N}(x_i - \overline{x})^2}$$

where σ = standard deviation

N = the number of measurements

x_i = the individual measured values

$\overline{x}$ = the average value of the measured quantity

To obtain a reliable estimate of standard deviation, N, the number of samples, should be fairly large. We can use statistical methods to determine a confidence interval on our measurements. A typical confidence level for scientific investigations is 90% or 95%.

Often in scientific operations we want to determine a quantity that requires many steps to measure. Of course, each time we take a measurement there will be a certain associated error that is a function of the measuring device. Each of these errors contributes to an even greater one in the final value. This phenomenon is known as **propagation of error** or propagation of uncertainty.

A measured value is typically expressed in the form x±Δx, where Δx is the uncertainty or margin of error. What this means is that the value of the measured quantity lies somewhere between x-Δx and x+Δx, but our measurement techniques do not allow us any more precision. If several measurements are required to ultimately decide a value, we must use formulas to determine the total uncertainty that results from all the measurement errors. A few of these formulas for simple functions are listed below:

Formula	Uncertainty
$X = A \pm B$	$(\Delta X)^2 = (\Delta A)^2 + (\Delta B)^2$
$X = cA$	$\Delta X = c\Delta A$
$X = c(A \cdot B)$	$\left(\dfrac{\Delta X}{X}\right)^2 = \left(\dfrac{\Delta A}{A}\right)^2 + \left(\dfrac{\Delta B}{B}\right)^2$
$X = c\left(\dfrac{A}{B}\right)$	$\left(\dfrac{\Delta X}{X}\right)^2 = \left(\dfrac{\Delta A}{A}\right)^2 + \left(\dfrac{\Delta B}{B}\right)^2$

For example, if we wanted to determine the density of a small piece of metal we would have to measure its weight on a scale and then determine its volume by measuring the amount of water it displaces in a graduated cylinder. There will be error associated with measurements made by both the scale and the graduated cylinder. Let's suppose we took the following measurements:

Mass: 57± 0.5 grams
Volume: 23 ± 3 mm^3

Since density is simply mass divided by the volume, we can determine its value to be:

$$\rho = \frac{m}{V} = \frac{57g}{23mm^3} = 2.5\frac{g}{mm^3}$$

Now we must calculate the uncertainty on this measurement, using the formula above:

$$\left(\frac{\Delta x}{x}\right)^2 = \left(\frac{\Delta A}{A}\right)^2 + \left(\frac{\Delta B}{B}\right)^2$$

$$\Delta x = \left(\sqrt{\left(\frac{\Delta A}{A}\right)^2 + \left(\frac{\Delta B}{B}\right)^2}\right)x = \left(\sqrt{\left(\frac{0.5g}{57g}\right)^2 + \left(\frac{3mm^3}{23mm^3}\right)^2}\right) \times 2.5\frac{g}{mm^3} = 0.3\frac{g}{mm^3}$$

Thus, the final value for the density of this object is 2.5 ± 0.3 g/mm^3.

Skill 4.5   Apply dimensional analysis and give an example of deriving an equation

Dimensional analysis is simply a technique in which the units of the variables in an equation are analyzed. It is often used by scientists and engineers to determine if a derived equation or computation is plausible. While it does not guarantee that a stated relationship is correct, it does tell us that the relationship is at least reasonable. When there is a mix of various physical quantities being equated in a relationship, the units of both sides of the equation must be the same. We will examine a simple example to demonstrate how dimensional analysis can be used:

The Ideal Gas Law is used to predict change in temperature, volume, or pressure of a gas. It is:

$$PV = nRT$$

Where P=pressure [Pa]
V=volume [m^3]
n=numbers of moles of gas [mol]
T=temperature [K]
R=the gas constant 8.314472 [m^3·Pa·K^{-1}·mol^{-1}]

Show that this formula is physically plausible using dimensional analysis.

Begin by substituting the units of each quantity into the equation:

$$[Pa] \times [m^3] = [mol] \left[\frac{[m^3] \times [Pa]}{[K] \times [mol]} \right] [K]$$

On the right-hand side of the equation, we can cancel [mol] and [K]:

$$[Pa] \times [m^3] = [m^3] \times [Pa]$$

Because this is a simple example, we can already see that the two sides of the equations have the same units. In more complicated equations, additional manipulation may be required to elucidate this fact. Now, we can confidently believe in the plausibility of this relationship.

Skill 4.6   Using mathematics to derive and solve equations

Mathematics is a very broad field, encompassing various specific disciplines including calculus, trigonometry, algebra, geometry, complex analysis and other areas. **Many physical phenomena can be modeled mathematically using functions to relate a specific parameter or state of the system to one or more other parameters**. For example, the net force on a charged object is a function of the direction and magnitude of the forces acting upon it, such as electrical attraction or repulsion, tension from a string or spring, gravity and friction. When a phenomenon has been modeled mathematically as a set of expressions or equations, the general relationships of numbers can be applied to glean further information about the system for the purposes of greater understanding or prediction of future behavior.

The process of treating a system mathematically can involve use of empirical relationships (such as Ohm's law) or more *a priori* relationships (such as the Schrödinger equation). Given or empirically derived equations can be manipulated or combined, depending on the specific situation, to isolate a specific parameter as a function of other parameters (solution of the equation). This process can also lead to a different equation that presents a new or simplified relationship among specific parameters (derivation of an equation).

Both derivation and solution of equations can be pursued in either an **exact** or **approximate** manner. **In some cases, equations are intractable and certain assumptions must be made to facilitate finding a solution**. These assumptions are typically drawn from generalizations concerning the behavior of a particular system and will result in certain restrictions on the validity and applicability of a solution. An exact solution, presumably, has none of these limitations.

Another variation of approximate derivation and solution of an equation involves numerical techniques. Ideally, an analytical approach that employs no approximations is the best alternative; such an approach yields the broadest and most useful results. Nevertheless, the intractability of an equation can lead to the need for either an approximate or numerical solution. One particular example is from the field of electromagnetics, where determination of the scattering of radiation from a finite circular cylinder is either extremely difficult or impossible to perform analytically. Thus, either numerical or approximate approaches are required. The range of numerical techniques available depends largely on the type of problem involved. The approach for numerically solving simple algebraic equations, for example, is different from the approach for numerically solving complex integral equations.

Care must be taken with analytical (and numerical or approximate) approaches to physical situations **as it is possible to produce mathematically valid solutions that are physically unacceptable.** For instance, the equations that describe the wave patterns produced by a disturbance of the surface of water, upon solution, may yield an expression that includes both incoming and outgoing waves. Nevertheless, in some situations, there is no physical source of incoming waves; thus, the analytical solution must be tempered by the physical constraints of the problem. In this case, one of the solutions, although mathematically valid, must be rejected as unphysical.

Skill 4.7 Applying trigonometric functions and graphing to solve problems (including vector problems)

Vector space is a collection of objects that have magnitude and direction. They may have mathematical operations, such as addition, subtraction, and scaling, applied to them. Vectors are usually displayed in boldface or with an arrow above the letter. They are usually shown in graphs or other diagrams as arrows. The length of the arrow represents the magnitude of the vector while the direction in which it points shows the direction.

To add two vectors graphically, the base of the second vector is drawn from the point of the first vector as shown below with vectors **A** and **B**. The sum of the vectors is drawn as a dashed line, from the base of the first vector to the tip of the second. As illustrated, the order in which the vectors are connected is not significant as the endpoint is the same graphically whether **A** connects to **B** or **B** connects to **A**. This principle is sometimes called the parallelogram rule.

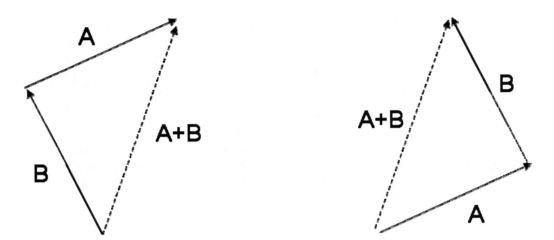

If more than two vectors are to be combined, additional vectors are simply drawn in accordingly with the sum vector connecting the base of the first to the tip of the final vector.

Subtraction of two vectors can be geometrically defined as follows. To subtract **A** from **B**, place the ends of **A** and **B** at the same point and then draw an arrow from the tip of **A** to the tip of **B**. That arrow represents the vector **B-A**, as illustrated below:

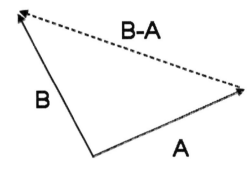

To add two vectors without drawing them, the vectors must be broken down into their orthogonal components using trigonometric functions. Add both x components to get the total x component of the sum vector, then add both y components to get the y component of the sum vector. Use the Pythagorean Theorem and the three trigonometric functions to the get the size and direction of the final vector.

Example: Here is a diagram showing the x and y-components of a vector D1:

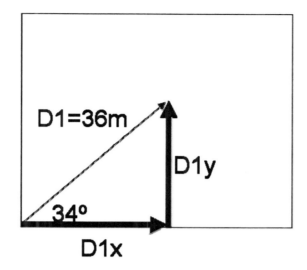

Notice that the x-component D1x is adjacent to the angle of 34 degrees.

Thus D1x=36m (cos34) =29.8m

The y-component is opposite to the angle of 34 degrees.

Thus D1y =36m (sin34) = 20.1m

A second vector D2 is broken up into its components in the diagram below using the same techniques. We find that D2y=9.0m and D2x=-18.5m.

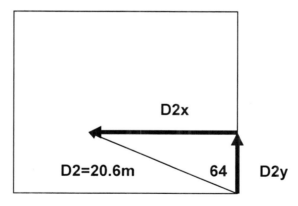

Next we add the x components and the y components to get

DTotal x =11.3 m   *and*   DTotal y =29.1 m

Now we have to use the Pythagorean theorem to get the total magnitude of the final vector. And the arctangent function to find the direction. As shown in the diagram below.

DTotal=31.2m

tan θ= DTotal y / DTotal x = 29.1m / 11.3 =2.6   θ=69 degrees

Skill 4.8   Using fundamental concepts of calculus to model and solve problems

The widespread use of calculus, as applied to problems of physics, can be largely credited to Sir Isaac Newton. Gottfried Wilhelm Leibniz is credited with having simultaneously developed calculus although his work did not have as much physics-oriented character as did Newton's work. The integral and differential, though abstract in and of themselves, can be related to various physical situations and can provide extremely useful analytical tools.

In simple terms, **the derivative of a function is essentially a slope or a rate of change of that function at each point in a defined range**. Thus, the derivative of a function that relates the position of a particle to time is the rate of change of the position, or speed. The derivative of a function over some range is itself a function, and can also be differentiated. Following the previous example, the derivative of the speed of a particle is the rate of change of the speed, or the acceleration. Thus, derivatives can be used to determine additional information about a system based on other information or observations.

In cases where a function involves more than one variable, **partial derivatives** may be involved. These are simply treated as derivatives in terms of the specified variable while assuming that all other variables remain constant. **Partial derivatives are a crucial component of operators such as the gradient, which, in part, determines the direction (in multiple dimensions) of the greatest increase of a function.**

The integral, also aptly named the antiderivative, is the area under the curve of a function in a given range. The function for the speed of a car can be integrated over time to get distance. This is a generalization of the case of constant speed where the speed is simply multiplied by the time traveled to get the distance traveled. The integral performs this multiplication operation over infinitesimally small increments of time thus allowing for calculation in the case of time-varying speed. Integrals may be performed for a single variable even in the case of a function containing several variables. The approach to integration in this case is the same as the approach to partial derivatives: all variables not being integrated are treated as constant.

Vector calculus is often necessary for physical situations where the directions of certain parameters, in addition to the magnitudes, are involved. Disciplines such as classical electrodynamics, for example, must largely be treated with vector calculus. For vector calculus, the choice of coordinate system becomes slightly more complicated, as in some cases the unit vectors are variable. The rectangular coordinate system involves unit vectors that are constant, making it a preferred system of coordinates so long as the expressions are not overly difficult. In these cases it may be beneficial to use other coordinate systems such as cylindrical or spherical.

COMPETENCY 5.0 **UNDERSTAND THE INTERRELATIONSHIPS AMONG PHYSICS, SOCIETY, TECHNOLOGY, AND OTHER SCIENCES AND DISCIPLINES**

Skill 5.1 **Explore the impact of physics and technology on society**

See **Skill 1.4** for information on this topic.

Skill 5.2 **Similarities and differences between science and technology (e.g., science as investigating the natural world, technology as solving human adaptation problems)**

The union of science, technology, and mathematics has shaped the world we live in today. Science describes the world. It attempts to explain all aspects of how nature works, from our own bodies to the tiny particles making up matter, from the entire earth to the universe beyond. Science lets us know in advance what will happen when a cell splits or when two chemicals react. Yet, science is ever-evolving. Throughout history, people have developed and validated many different ideas about the processes of the universe. Frequently, the development of new technology used in conducting experiments allows for new information and theories to emerge.

Technology makes use of scientific knowledge to solve real-world problems. For example, science is used to study the flow of electrons but technology is required to channel the flow of electrons to create a supercomputer. **Basic research** generally refers to investigation of fundamental scientific principles. **Applied research** is oriented toward making use of basic research in technology development. Applied research is dependent on basic research, and both are necessary for technology advancement. Mathematics in turn provides the language that allows this knowledge to be communicated. It allows the creation of models for scientists to use in explaining natural phenomena and is also the language of technology and computers.

The economy today is dependent upon the existence of technology. The job market changes as new technologies develop. For example, with the advent of computers, many trained workers in information science and technology are needed, moving our economy from a manufacturing-based economy to a knowledge-based economy. Industrial labs are being redefined or eliminated, creating a convergence between scientific disciplines and engineering and providing new entrepreneurial opportunities.

At the same time, advances in scientific knowledge and technology often present ethical dilemmas for society. Industrialization brought the consumption of great amounts of energy. This in turned creates environmental problems, contributes to global conflicts, and depletes natural resources. Scientific knowledge tells us there is oil as shale that we have not yet accessed. Technology developments may allow us to cost-effectively reach this oil and power our economy. At the same time, new technologies will emerge providing for alternative power supplies, such as wind farms or biofuels. New types of skills will be needed to support these new technologies and the job market; education and our society will shift in response.

Skill 5.3   Understand the technological design process

Technology can include a broad range of human designs ranging from pharmaceuticals to manufacturing processes to simple tools. Thus the exact process by which new technologies are designed varies across different fields but the general outline is as follows:

Identifying need
In this first step in developing a new technology, a human need which science can help meet is identified. This need may be general or specific, an ancient problem or a new one. Sometimes, new discoveries or developments cause scientists to consider what problems might be solved by making use of the new information or technology.

Researching and specifying constraints/ functions
During this phase, developers learn all they can about the problem they are hoping to address. This may also include studying previous technologies and their advantages and limitations. They will make a complete study of the problem and all its aspects. Ideally, developers will arrive at a function list for their new technology - a list of the capabilities the new technology must possess along with any known constraints.

Conceptualizing
The initial design for the new technology is developed during this phase. Some small models and drawings may be made which help to visualize and refine the proposed design. Choices (appropriate materials, instrumentation, etc.) will be made to ensure that the function list is satisfied. Developers try to be very specific about their plans for their final design but still allow for some flexibility and changes as the design will likely change in the next phase of development.

Prototyping and Testing
Models of the new technology are created during this step of the process. These prototypes may be smaller or simpler than the envisioned final design or may be very similar to it. The prototypes are tested in a variety of ways to help assess how well the design will work in its final environment. This phase could include everything from testing new medical products in animal models to running time trials with new automobiles.

Finalizing
During this phase, the final form of the technology is tested in its real world application. Some additional testing and small refinements may be necessary. If the technology is a new product, preparations may be made for its manufacture and/or for developing packaging. Regulatory specifications may also need to be satisfied. For most cases, redesign continues as the technology evolves.

Skill 5.4   Explore some ethical considerations related to science and technology

Science and technology are often referred to as a "double- edged sword". Although advances in medicine have greatly improved the quality and length of life, certain moral and ethical controversies have arisen. Unforeseen environmental problems may result from technological advances. Advances in science have led to an improved economy through biotechnology as applied to agriculture, yet it has put our health care system at risk and has caused the cost of medical care to skyrocket. Society depends on science, yet is necessary that the public be scientifically literate and informed in order to prevent potentially unethical procedures from occurring. Especially vulnerable are the areas of genetic research and fertility. It is important for science teachers to stay abreast of current research and to involve students in critical thinking and ethics whenever possible.

See the next section, **Skill 5.5**, for further exploration of this topic.

Skill 5.5   Application of scientific and technological decision making at the community, state, national, and international level

Advances in science and technology create challenges and ethical dilemmas that national governments and society in general must attempt to solve. Local, state, national, and global governments and organizations must increasingly consider policy issues related to science and technology. For example, local and state governments must analyze the impact of proposed development and growth on the environment. Governments and communities must balance the demands of an expanding human population with the local ecology to ensure sustainable growth. Genetic research and manipulation, antibiotic resistance, stem cell research, and cloning are but a few of the issues facing national governments and global organizations today.

In all cases, policy makers must analyze all sides of an issue and attempt to find a solution that protects society while limiting scientific inquiry as little as possible. For example, policy makers must weigh the potential benefits of stem cell research, genetic engineering, and cloning (e.g. medical treatments) against the ethical and scientific concerns surrounding these practices. Many safety concerns have answered by strict government regulations. The FDA, USDA, EPA, and National Institutes of Health are just a few of the government agencies that regulate pharmaceutical, food, and environmental technology advancements

Scientific and technological breakthroughs greatly influence other fields of study and the job market as well. Advances in information technology have made it possible for all academic disciplines to utilize computers and the internet to simplify research and information sharing. In addition, science and technology influence the types of available jobs and the desired work skills. For example, machines and computers continue to replace unskilled laborers and computer and technological literacy is now a requirement for many jobs and careers. Finally, science and technology continue to change the very nature of careers. Because of science and technology's great influence on all areas of the economy, and the continuing scientific and technological breakthroughs, careers are far less stable than in past eras. Workers can thus expect to change jobs and companies much more often than in the past.

Because people often attempt to use scientific evidence in support of political or personal agendas, the ability to evaluate the credibility of scientific claims is a necessary skill in today's society. The media and those with an agenda to advance often overemphasize the certainty and importance of experimental results. One should question any scientific claim that sounds fantastical or overly certain. Scientific, peer-reviewed journals are the most accepted source for information on scientific experiments and studies. Knowledge of experimental design and the scientific method is important in evaluating the credibility of studies. For example, one should look for the inclusion of control groups and the presence of data to support the given conclusions.

COMPETENCY 6.0   UNDERSTAND SAFE AND PROPER USE OF
EQUIPMENT, MATERIALS, AND CHEMICALS USED
IN PHYSICS INVESTIGATIONS

Skill 6.1   Explain the principles upon which given laboratory
instruments are based (e.g., oscilloscopes, voltmeter, Geiger
counters)

Oscilloscope: An oscilloscope is a piece of electrical test equipment that allows
signal voltages to be viewed as two-dimensional graphs of electrical potential
differences plotted as a function of time.

The oscilloscope functions by measuring the deflection of a beam of electrons
traveling through a vacuum in a cathode ray tube. The deflection of the beam can
be caused by a magnetic field outside the tube or by electrostatic energy created
by plates inside the tube. The unknown voltage or potential energy difference
can be determined by comparing the electron deflection it causes to the electron
deflection caused by a known voltage.

Oscilloscopes can also determine if an electrical circuit is oscillating and at what
frequency. They are particularly useful for troubleshooting malfunctioning
equipment. You can see the "moving parts" of the circuit and tell if the signal is
being distorted. With the aid of an oscilloscope you can also calculate the
"noise" within a signal and see if the "noise" changes over time.

Inputs of the electrical signal are usually entered into the oscilloscope via a
coaxial cable or probes. A variety of transducers can be used with an
oscilloscope that enable it to measure other stimuli including sound, pressure,
heat, and light.

Voltmeter/Ohmmeter/Ammeter: A common electrical meter, typically known as
a multimeter, is capable of measuring voltage, resistance, and current. Many of
these devices can also measure capacitance (farads), frequency (hertz), duty
cycle (a percentage), temperature (degrees), conductance (siemens), and
inductance (henrys).

These meters function by utilizing the following familiar equations:

Across a resistor (Resistor R):

$$V_R = IR_R$$

Across a capacitor (Capacitor C):

$$V_C = IX_C$$

Across an inductor (Inductor L):

$$V_L = IX_L$$

Where V=voltage, I=current, R=resistance, X=reactance.

If any two factors in the equations are held constant or are known, the third factor can be determined and is displayed by the multimeter.

Signal Generator: A signal generator, also known as a test signal generator, function generator, tone generator, arbitrary waveform generator, or frequency generator, is a device that generates repeating electronic signals in either the analog or digital domains. They are generally used in designing, testing, troubleshooting, and repairing electronic devices.

A function generator produces simple repetitive waveforms by utilizing a circuit called an electronic oscillator or a digital signal processor to synthesize a waveform. Common waveforms are sine, sawtooth, step or pulse, square, and triangular. Arbitrary waveform generators are also available which allow a user to create waveforms of any type within the frequency, accuracy and output limits of the generator. Function generators are typically used in simple electronics repair and design where they are used to stimulate a circuit under test. A device such as an oscilloscope is then used to measure the circuit's output.

Spectrometer: A spectrometer is an optical instrument used to measure properties of light over a portion of the electromagnetic spectrum. Light intensity is the variable that is most commonly measured but wavelength and polarization state can also be determined. A spectrometer is used in spectroscopy for producing spectral lines and measuring their wavelengths and intensities. Spectrometers are capable of operating over a wide range of wavelengths, from short wave gamma and X-rays into the far infrared. In optics, a spectrograph separates incoming light according to its wavelength and records the resulting spectrum in some detector. In astronomy, spectrographs are widely used with telescopes.

Geiger counter: A Geiger counter detects ionizing radiation, i.e. radiation capable of ionizing atoms and molecules and causing damage to DNA and living tissue. The core of a Geiger counter is a Geiger-Müller tube, a tube filled with inert gas (typically helium, neon or argon) and containing a thin metal wire carrying a charge. When a nuclear particle penetrates the tube, it temporarily makes the gas conductive and electrons are ultimately attracted to the central conducting wire. As the particle travels towards the wire, it continues to collide with atoms and release more electrons, thereby amplifying the signal. Ultimately, a pulse is produced by each particle that enters the detector. The counters are designed to display this signal with a needle or lamp or to produce audible clicks. The earliest device of this type was produced by Ernest Rutherford in 1908 and the design was later refined by Hans Geiger and Walther Müller 20 years later. They are presently called "counters" because every particle produces a pulse or click, allowing the total number of particles to be tabulated. Geiger counters are commonly used because they are sturdy and relatively cheap to produce. However, they are only capable of determining the intensity of radiation (number of particles present over a given time). The energy level of the particles is also an important quantity but basic Geiger counters are not able to measure this.

Skill 6.2 Give examples of some hazards associated with given laboratory materials (e.g., projectiles, lasers, radioactive materials, heat sources)

Laboratory and field equipment used for scientific investigation must be handled with the greatest caution and care. The teacher must be completely familiar with the use and maintenance of a piece of equipment before it is introduced to the students. Maintenance procedures for equipment must be scheduled and recorded and each instrument must be calibrated and used strictly in accordance with the specified guidelines in the accompanying manual. Following are some safety precautions one can take in working with different types of equipment:

1. Electricity: Safety in this area starts with locating the main cut off switch. All the power points, switches, and electrical connections must be checked one by one. Batteries and live wires must be checked. All checking must be done with the power turned off. The last act of assembling is to insert the plug and the first act of disassembling is to take off the plug.

2. Motion and forces: All stationary devices must be secured by C-clamps. Protective goggles must be used. Care must be taken at all times while knives, glass rods and heavy weights are used. Viewing a solar eclipse must always be indirect. When using model rockets, NASA's safety code must be implemented.

3. Heat: The master gas valve must be off at all times except while in use. Goggles and insulated gloves are to be used whenever needed. Never use closed containers for heating. Burners and gas connections must be checked periodically. Gas jets must be closed soon after the experiment is over. Fire retardant pads and quality glassware such as Pyrex must be used.

4. Pressure: While using a pressure cooker, never allow pressure to exceed 20 lb/square inch. The pressure cooker must be cooled before it is opened. Care must be taken when using mercury since it is poisonous. A drop of oil on mercury will prevent the mercury vapors from escaping.

5. Light: Broken mirrors or those with jagged edges must be discarded immediately. Sharp-edged mirrors must be taped. Spectroscopic light voltage connections must be checked periodically. Care must be taken while using ultraviolet light sources. Some students may have psychological or physiological reactions to the effects of strobe like (e.g. epilepsy).

6. Lasers: Direct exposure to lasers must not be permitted. The laser target must be made of non-reflecting material. The movement of students must be restricted during experiments with lasers. A number of precautions while using lasers must be taken – use of low power lasers, use of approved laser goggles, maintaining the room's brightness so that the pupils of the eyes remain small. Appropriate beam stops must be set up to terminate the laser beam when needed. Prisms should be set up before class to avoid unexpected reflection.

7. Sound: Fastening of the safety disc while using the high speed siren disc is very important. Teacher must be aware of the fact that sounds higher than 110 decibels will cause damage to hearing.

8. Radiation: Proper shielding must be used while doing experiments with x-rays. All tubes that are used in a physics laboratory such as vacuum tubes, heat effect tubes, magnetic or deflection tubes must be checked and used for demonstrations by the teacher. Cathode rays must be enclosed in a frame and only the teacher should move them from their storage space. Students must watch the demonstration from at least eight feet away.

9. Radioactivity: The teacher must be knowledgeable and properly trained to handle the equipment and to demonstrate. Proper shielding of radioactive material and proper handling of material are absolutely critical. Disposal of any radioactive material must comply with the guidelines of NRC.

Skill 6.3   Outline some basic safety rules for electricity and electrical equipment

Active learning about electromagnetism may require the use of live electrical sources and many physics laboratories can be greatly enhanced with the use of electrical devices. However, care must always be taken when dealing with electricity. The danger in working with electrical device is two-fold as follows.

1. Risk of electrical shock due to:
    a. Exposed electrical contacts
    b. Poor or damaged insulation in equipment or on cords
    c. Moisture in the area of electrical equipment

2. Risk of fire due to:
    a. Over-heating of equipment or overloaded circuits
    b. Faulty connections or short circuits
    c. Flammable substance in the area of electrical equipment

Adherence to the following rules will minimize the risk of electrical accidents:
1. Experiments should never be performed by a single person working alone.
2. Check all equipment, switches, and leads one by one. If missing insulation, frayed cords, blackening (due to arcing), bent or missing prongs are in evidence, do not use the equipment.
3. Always connect and turn the power supply on last. When concluding work, turn the power supply off first.
4. Do not run wires (or extension cords) over moving or rotating equipment, or on the floor, or string them across walkways from bench-to-bench.
5. Do not wear conductive watch bands or chains, finger rings, wrist watches, etc., and do not use metallic pencils, metal or metal edge rulers when working with exposed circuits.
6. If breaking an inductive circuit open, turn your face away to avoid danger from possible arcing.
7. If using electrolytic capacitors, always wait an appropriate period of time (usually approximately five time constants) for capacitor discharge before working on the circuit.
8. All conducting surfaces serving as "ground" should be connected together.
9. All instructional laboratories should be equipped with Ground Fault Current Interrupt (GFCI) circuit breakers. If breakers repeatedly trip and overload is not present, check for leakage paths to ground.
10. Use only dry hands and tools, work on a dry surface, and stand on a dry surface.
11. Never put conductive metal objects into energized equipment.
12. Always carry equipment by the handle and/or base, never by the cord.
13. Unplug cords by pulling on the plug, never on the cord.
14. Use extension cords only temporarily. The cord should be appropriately rated for the job.

15. Use extension cords with 3 prong plugs to ensure that equipment is grounded.
16. Never remove the grounding post from a 3 prong plug.
17. Do not overload extension cords, multi-outlet strips, or wall outlets.

Skill 6.4   Demonstrate knowledge of proper procedures for dealing with accidents and injuries in the physics laboratory

Safety is a learned behavior and must be incorporated into instructional plans. Measures of prevention and procedures for dealing with emergencies in hazardous situations have to be in place and readily available for reference. Copies of these must be given to all people concerned, such as administrators and students.

The single most important aspect of safety is planning and anticipating various possibilities and preparing for the eventuality. Any Physics teacher/educator planning on doing an experiment must try it before the students do it. In the event of an emergency, quick action can prevent many disasters. The teacher/educator must be willing to seek help at once without any hesitation because sometimes it may not be clear that the situation is hazardous and potentially dangerous. Accidents and injuries should always be reported to the school administration and local health facilities. The severity of the accident or injury will determine the course of action to pursue.

There are a number of procedures to prevent and correct any hazardous situation. There are several safety aids available commercially such as posters, safety contracts, safety tests, safety citations, texts on safety in secondary classroom/laboratories, hand books on safety and a host of other equipment. Another important thing is to check the laboratory and classroom for safety and report it to the administrators before staring activities/experiments.

SUBAREA II   **MECHANICS AND HEAT ENERGY**

COMPETENCY 7.0   **UNDERSTAND CONCEPTS RELATED TO MOTION IN ONE AND TWO DIMENSIONS, AND APPLY THIS KNOWLEDGE TO SOLVE PROBLEMS THAT REQUIRE THE USE OF ALGEBRA, CALCULUS, AND GRAPHING**

Skill 7.1   **Explain and give examples of the terminology, units, and equations used to describe and analyze one and two-dimensional motion**

Kinematics is the part of mechanics that seeks to understand the motion of objects, particularly the relationship between position, velocity, acceleration and time.

The above figure represents an object and its displacement along one linear dimension.

First we will define the relevant terms:

1. Position or Distance is usually represented by the variable x. It is measured relative to some fixed point or datum called the origin in linear units, meters, for example.

2. Displacement is defined as the change in position or distance which an object has moved and is represented by the variables D, d or Δx. Displacement is a vector with a magnitude and a direction.

3. Velocity is a vector quantity usually denoted with a V or v and defined as the rate of change of position. Typically units are distance/time, m/s for example. Since velocity is a vector, if an object changes the direction in which it is moving it changes its velocity even if the speed (the scalar quantity that is the magnitude of the velocity vector) remains unchanged.

i) Average velocity: $\vec{v} = \frac{\Delta d}{\Delta t} = d_1 - d_0 / t_1 - t_0$.

The ratio, $\Delta d / \Delta t$ is called the average velocity. Average here denotes that this quantity is defined over a period Δt.

ii) Instantaneous velocity is the velocity of an object at a particular moment in time. Conceptually, this can be imagined as the extreme case when Δt is infinitely small.

5. Acceleration represented by a is defined as the rate of change of velocity and the units are m/s^2. Both an average and an instantaneous acceleration can be defined similarly to velocity.

From these definitions we develop the kinematic equations. In the following, subscript i denotes initial and subscript f denotes final values for a time period. Acceleration is assumed to be constant with time.

$$v_f = v_i + at \qquad (1)$$

$$d = v_i t + \frac{1}{2}at^2 \qquad (2)$$

$$v_f^2 = v_i^2 + 2ad \qquad (3)$$

$$d = \left(\frac{v_i + v_f}{2}\right)t \qquad (4)$$

The same relationships apply between distance, velocity, acceleration, and time in two dimensions as apply in one dimension, but each dimension must be treated separately. For a discussion of projectile motion in two dimensions see **Skill 7.2**.

Motion on an arc can also be considered from the view point of the kinematic equations. As pointed out earlier, displacement, velocity and acceleration are all vector quantities, i.e. they have magnitude (the speed is the magnitude of the velocity vector) and direction. This means that if one drives in a circle at constant speed one still experiences an acceleration that changes the direction. We can define a couple of parameters for objects moving on circular paths and see how they relate to the kinematic equations.

1. Tangential speed: The tangent to a circle or arc is a line that intersects the arc at exactly one point. If you were driving in a circle and instantaneously moved the steering wheel back to straight, the line you would follow would be the tangent to the circle at the point where you moved the wheel. The tangential speed then is the instantaneous magnitude of the velocity vector as one moves around the circle.

2. Tangential acceleration: The tangential acceleration is the component of acceleration that would change the tangential speed and this can be treated as a linear acceleration if one imagines that the circular path is unrolled and made linear.

3. Centripetal acceleration: Centripetal acceleration corresponds to the constant change in the direction of the velocity vector necessary to maintain a circular path. Always acting toward the center of the circle, centripetal acceleration has a magnitude proportional to the tangential speed squared divided by the radius of the path.

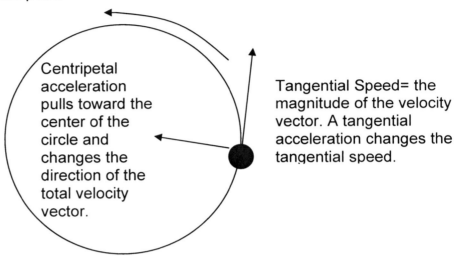

Centripetal acceleration pulls toward the center of the circle and changes the direction of the total velocity vector.

Tangential Speed= the magnitude of the velocity vector. A tangential acceleration changes the tangential speed.

Skill 7.2 Analyze the movement of freely falling objects near the surface of the earth

The most common example of an object moving near the surface of the earth is a projectile. A projectile is an object upon which the only force acting is gravity.
Some examples:
i) An object dropped from rest
ii) An object thrown vertically upwards at an angle
iii) A canon ball

Once a projectile has been put in motion (say, by a canon or hand) the only force acting on it is gravity which near the surface of the earth is characterized by the acceleration $a=g=9.8 m/s^2$.

This is most easily considered with an example such as the case of a bullet shot horizontally from a standard height at the same moment that a bullet is dropped from exactly the same height. Which will hit the ground first? If we assume wind resistance is negligible, then the acceleration due to gravity is our only acceleration on either bullet and we must conclude that they will hit the ground at the same time. The horizontal motion of the bullet is not affected by the downward acceleration.

Example:
I shoot a projectile at 1000 m/s from a perfectly horizontal barrel exactly 1 m above the ground. How far does it travel before hitting the ground?

Solution:
First figure out how long it takes to hit the ground by analyzing the motion in the vertical direction. In the vertical direction, the initial velocity is zero so we can rearrange kinematic equation 2 from the previous section to give:

$t = \sqrt{\dfrac{2d}{a}}$. Since our displacement is 1 m and a=g=9.8m/s^2, t=0.45 s.

Now use the time to hitting the ground from the previous calculation to calculate how far it will travel horizontally. Here the velocity is 1000m/s and there is no acceleration. So we simple multiply velocity with time to get the distance of 450m.

Skill 7.3   Solve problems involving distance, displacement, speed, velocity, and constant acceleration

Simple problems involving distance, displacement, speed, velocity, and constant acceleration can be solved by applying the kinematics equations from a preceding section (**Skill 7.1**). The following steps should be employed to simplify a problem and apply the proper equations:

1. Create a simple diagram of the physical situation.
2. Ascribe a variable to each piece of information given.
3. List the unknown information in variable form.
4. Write down the relationships between variables in equation form.
5. Substitute known values into the equations and use algebra to solve for the unknowns.
6. Check your answer to ensure that it is reasonable.

Example:
A man in a truck is stopped at a traffic light. When the light turns green, he accelerates at a constant rate of 10 m/s^2. **a)** How fast is he going when he has gone 100 m? **b)** How fast is he going after 4 seconds? **c)** How far does he travel in 20s?

$a=10$ m/s^2

$v_i=0$ m/s

Solution:
We first construct a diagram of the situation.
In this example, the diagram is very simple, only showing the truck accelerating at the given rate. Next we define variables for the known quantities (these are noted in the diagram):
$$a=10 \text{ m/s}^2; \quad v_i=0 \text{ m/s}$$
Now we will analyze each part of the problem, continuing with the process outlined above.

For part **a)**, we have one additional known variable: d=100 m.
The unknowns are: v_f (the velocity after the truck has traveled 100m)
Equation (3) will allow us to solve for v_f, using the known variables:

$$v_f^2 = v_i^2 + 2ad$$

$$v_f^2 = (0m/s)^2 + 2(10m/s^2)(100m) = 2000\frac{m^2}{s^2}$$

$$v_f = 45\frac{m}{s}$$

We use this same process to solve part **b)**. We have one additional known variable: t=4 s. The unknowns are: v_f (the velocity after the truck has traveled for 4 seconds). Thus, we can use equation (1) to solve for v_f:

$$v_f = v_i + at$$

$$v_f = 0m/s + (10m/s^2)(4s) = 40m/s$$

For part **c)**, we have one additional known variable: t= 20 s.
The unknowns are: d (the distance after the truck has traveled for 20 seconds).
Equation (2) will allow us to solve this problem:

$$d = v_i t + \frac{1}{2}at^2$$

$$d = (0m/s)(20s) + \frac{1}{2}(10m/s^2)(20s)^2 = 2000m$$

Finally, we consider whether these solutions seem physically reasonable. In this simple problem, we can easily say that they do.

Skill 7.4   Interpret information presented in graphic representations related to displacement, velocity, and constant acceleration

Trends and patterns in a set of data are most easily identified when the data is displayed graphically. Exact data values in tabular form are also used to calculate various features of the data set. Following are some aspects of data that are commonly analyzed in all scientific disciplines.

The **slope** or the **gradient** of a line is used to describe the rate of change of a variable with respect to another or, in calculus terms, the **derivative** of one variable with respect to another. In the set of examples shown below, the relationship between time, position or distance, velocity and acceleration can be understood conceptually by looking at a graphical representation of each as a function of time. The velocity is the slope of the position vs. time graph and the acceleration is the slope of the velocity vs. time graph.

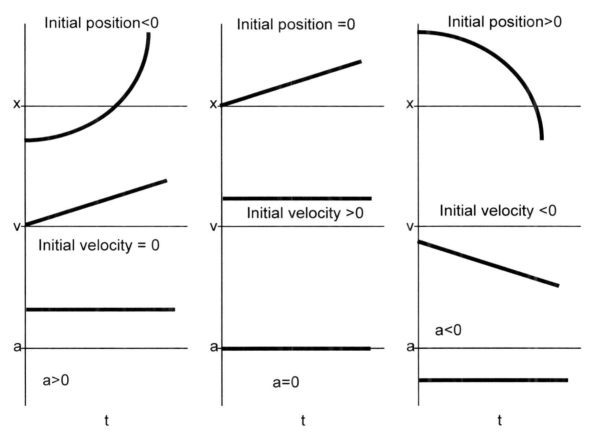

Here are some things we notice by inspecting these graphs:
1) In each case acceleration is constant.
2) A non-zero acceleration produces a position curve that is a parabola.
3) In each case the initial velocity and position are specified. The acceleration curve gives the shape of the velocity curve, but not the initial value and the velocity curve gives the shape of the position curve but not the initial position.

COMPETENCY 8.0   UNDERSTAND CHARACTERISTICS OF FORCES AND METHODS USED TO MEASURE FORCE, AND SOLVE ALGEBRAIC PROBLEMS INVOLVING FORCES

Skill 8.1 Identify forces acting in a given situation

Free body diagrams are simple sketches that show all the objects and forces in a given physical situation. This makes them very useful for understanding and solving physical problems. These diagrams show relative positions, masses, and the direction in which forces are acting. Some of the common forces that act on a body are discussed and illustrated here using free body diagrams.

Gravity
This is the force that pulls a body towards the center of the earth, i.e. downwards, and is also called the weight of the body. It is given by
$$W = mg$$
Where m is the mass of the body and $g=9.81$ m/s² is the acceleration due to gravity.

Normal force
When a body is pressed against a surface it experiences a reaction force that is perpendicular to the surface and in the direction away from the surface. For instance, an object resting on a table experiences an upward reaction force from the table that is equal and opposite to the force that the object exerts on the table. When the table is horizontal and no additional force is being applied to the object, the normal force is equal to the weight of the object.

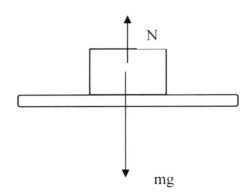

Friction

Friction is the force on a body that opposes its sliding over a surface. This force is due to the bonding between the two surfaces and is greater for rough surfaces. It acts in the direction opposite to the force attempting to move the object. When the object is at rest, the frictional force is known as **static friction**. The frictional force on an object in motion is known as **kinetic friction**.

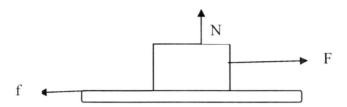

The frictional force is usually directly proportional to the normal force and can be calculated as $F_f = \mu\, F_n$ where μ is either the coefficient of static friction or kinetic friction depending on whether the object is at rest or in motion.

Tension/Compression

Tension is the force that acts in a rope, cable or rod that is attached to something and is being pulled. Tension acts along the cord. When a hand pulls a rope attached to a box, for instance, the tension T in the rope acts to pull the rope apart while it works on the box and the hand in the opposite direction as shown below:

Compression is the opposite of tension in that the force acts to shorten a rigid body instead of pulling it apart.

Pressure

For a description of fluid pressure and the force of buoyancy see **Skill 15.1.**

Net force

The forces that act on a body come from many different sources. Their effect on a body, however, is the same; a change in the state of motion of the body as given by Newton's laws of motion. Therefore, once we identify the magnitude and direction of each force acting on a body, we can combine the effect of all the forces together using vector addition and find the net force.

Problem:
Find the net force on a 5 Kg box sliding down an inclined surface at an angle of 30^0 with the horizontal if the coefficient of friction of the surface is 0.5.

Solution:
There are three forces acting on the box, gravity, the normal force and the frictional force. We can resolve these forces along the inclined plane and perpendicular to the plane and find the net force in each direction.

Perpendicular to the plane:
The component of the gravitational force perpendicular to the plane = mgcos30 = 5 x 9.8 x 0.87 = 42.6N
The normal force acting on the box is equal and opposite to the perpendicular component of the gravitational force. Thus the net force on the box perpendicular to the plane is zero.

Along the inclined plane:
The component of the gravitational force down the plane =mgsin30 = 5 x 9.8 x 0.5 = 24.5N
The force of friction up the plane = μ F_n = 0.5 x 42.6 = 21.3N
Thus the net force on the box acts down the plane and is equal to 24.5 – 21.3 = 3.2N

Skill 8.2   Experimental designs for measuring forces

Designing an experiment to accurately measure a force directly can be extremely challenging and indirect methods are often required. For example, to measure the force on a moving object, it may be simpler to measure the position of the object with time, differentiate the results twice to get the acceleration and then multiply by the mass of the object in accordance with F = ma. Thus, force measurement experiments may often be more complicated than simply using a mechanical scale delineated in Newtons.

Indirect force measurement through fields or potentials
Forces may be measured indirectly through a field or potential measurement. Since it is impossible to attach a tiny spring scale to an electron in order to measure the force of an electric field upon it, a measurement of the electric field or electric potential can be used instead. Such measurements may be taken by way of a field meter or multimeter, providing information about the field or potential from which the force on an electron (or other object) can be calculated.

Indirect force measurements through material characteristics

Properties of various materials can also be used to indirectly measure a force. The piezoelectric effect is one example that results from the ability of certain materials, such as some crystals and ceramics, to produce an electric potential when they are stressed mechanically. Thus, these natural "pressure sensors" can be used for force measurements through measurement of the voltage produced in a given situation. One particular application of piezoelectric materials is in equipment for fine weight measurements that can be used to measure the gravitational force on small or low-mass objects.

The temperature and volume of a confined gas may also be used to measure force by using, for example, a piston. Force applied to the piston causes an increase in the pressure of the gas and, by the ideal gas law ($PV = nRT$), a corresponding change in the volume and temperature of the gas. The pressure can be calculated from these parameters (or measured directly), thus allowing the calculation of the force on the piston.

Stages of experimental design

Experimental apparatuses for measurement of forces, therefore, involve several aspects. First, the transducer, which may be more or less a direct measurement of the force, acts as a converter of force into some intelligible signal. In some cases, as with the examples of the piston or the piezoelectric potential, the magnitude of the force must be calculated from the signal (whether electric or otherwise) through a theoretical relationship determined by the characteristics of the material or materials used. Second, for the more indirect force measurements, computer equipment may be required for calculating the force based on other parameters in the experiment.

In the example of the moving object, it is most convenient to use a computer for performing the differentials, although, for sufficiently small data sets, this can be done "by hand." Regardless of the complexity, an algorithm for calculating the force values must be followed, and this is as much part of the experimental apparatus as the transducer equipment. In simple cases, however, such as spring scales, the force can be measured by simply looking at the reading.

Skill 8.3   Solve problems involving gravitational and frictional forces

Newton's universal law of **gravitation** states that any two objects experience a force between them as the result of their masses. Specifically, the force between two masses m_1 and m_2 can be summarized as

$$F = G \frac{m_1 m_2}{r^2}$$

where G is the gravitational constant ($G = 6.672 \times 10^{-11} \, Nm^2 / kg^2$), and r is the distance between the two objects.

The weight of an object is the result of the gravitational force of the earth acting on its mass. The acceleration due to Earth's gravity on an object is 9.81 m/s². Since force equals mass * acceleration, the magnitude of the gravitational force created by the earth on an object is

$$F_{Gravity} = m_{object} \cdot 9.81 \, m / s^2$$

Important things to remember:

1. The gravitational force is proportional to the masses of the two objects, but *inversely* proportional to the *square of the distance* between the two objects.
2. When calculating the effects of the acceleration due to gravity for an object above the earth's surface, the distance above the surface is ignored because it is inconsequential compared to the radius of the earth. The constant figure of 9.81 m/s² is used instead.

Problem: Two identical 4 kg balls are floating in space, 2 meters apart. What is the magnitude of the gravitational force they exert on each other?

Solution:

$$F = G \frac{m_1 m_2}{r^2} = G \frac{4 \times 4}{2^2} = 4G = 2.67 \times 10^{-10} \, N$$

In the real world, whenever an object moves its motion is opposed by a force known as **friction.** How strong the frictional force is depends on numerous factors such as the roughness of the surfaces (for two objects sliding against each other) or the viscosity of the liquid an object is moving through. Most problems involving the effect of friction on motion deal with sliding friction. This is the type of friction that makes it harder to push a box across cement than across a marble floor.

When you try and push an object from rest, you must overcome the maximum **static friction** force to get it to move. Once the object is in motion, you are working against **kinetic friction** which is smaller than the static friction force previously mentioned. Sliding friction is primarily dependent on two things, the **coefficient of friction (μ)** which is primarily dependent on roughness of the surfaces involved and the amount of force pushing the two surfaces together. This force is also known as the **normal force (F_n)**, the perpendicular force between two surfaces. When an object is resting on a flat surface, the normal force is pushing opposite to the gravitational force – straight up. When the object is resting on an incline, the normal force is less (because it is only opposing that portion of the gravitational force acting perpendicularly to the object) and its direction is perpendicular to the surface of incline but at an angle from the ground. Therefore, for an object experiencing no external action, the magnitude of the normal force is either equal to or less than the magnitude of the gravitational force (F_g) acting on it. The frictional force (F_f) acts perpendicularly to the normal force, opposing the direction of the object's motion.

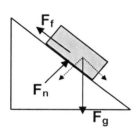

The frictional force is normally directly proportional to the normal force and, unless you are told otherwise, can be calculated as $F_f = \mu\, F_n$ where μ is either the coefficient of static friction or kinetic friction depending on whether the object starts at rest or in motion. In the first case, the problem is often stated as "how much force does it take to start an object moving" and the frictional force is given by $F_f > \mu_s\, F_n$ where μ_s is the coefficient of static friction. When questions are of the form "what is the magnitude of the frictional force opposing the motion of this object," the frictional force is given by $F_f = \mu_k\, F_n$ where μ_k is the coefficient of kinetic friction. There are several important things to remember when solving problems about friction.

1. The frictional force acts in opposition to the direction of motion.
2. The frictional force is proportional, and acts perpendicular to, the normal force.
3. The normal force is perpendicular to the surface the object is lying on. If there is a force pushing the object against the surface, it will increase the normal force.

Problem:
A woman is pushing an 800N box across the floor. She pushes with a force of 1000 N. The coefficient of kinetic friction is 0.50. If the box is already moving, what is the force of friction acting on the box?

Solution:
First it is necessary to solve for the normal force.
F_n= 800N + 1000N (sin 30°) = 1300N
Then, since $F_f = \mu\, F_n$ = 0.5*1300=650N

COMPETENCY 9.0 **APPLY KNOWLEDGE OF VECTORS AND TRIGONOMETRIC FUNCTIONS TO SOLVE PROBLEMS INVOLVING CONCURRENT, PARALLEL, RESULTANT, EQUILIBRANT, AND COMPONENT FORCES AND TORQUE**

Skill 9.1 **Apply graphic solutions to solve problems involving concurrent and equilibrant forces**

An object is said to be in a state of equilibrium when the forces exerted upon it are balanced. That is to say, forces to the left balance the forces exerted to the right, and upward forces are balanced by downward forces. The net force acting on the object is zero and the acceleration is 0 meters per second squared. This does not necessarily mean that the object is at rest. According to Newton's first law of motion, an object at equilibrium is either at rest and remaining at rest (**static equilibrium**), or in motion and continuing in motion with the same speed and direction (**dynamic equilibrium**).

Equilibrium of forces is often used to analyze situations where objects are in static equilibrium. One can determine the weight of an object in static equilibrium or the forces necessary to hold an object at equilibrium. The following are examples of each type of problem.

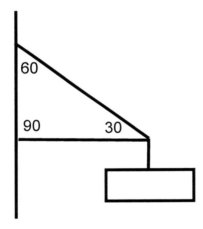

Problem: A sign hangs outside a building supported as shown in the diagram. The sign has a mass of 50 kg. Calculate the tension in the cable.

Solution: Since there is only one upward pulling cable it must balance the weight. The sign exerts a downward force of 490 N. Therefore, the cable pulls upwards with a force of 490 N. It does so at an angle of 30 degrees. To find the total tension in the cable:

$$F_{total} = 490 \text{ N} / \sin 30°$$
$$F_{total} = 980 \text{ N}$$

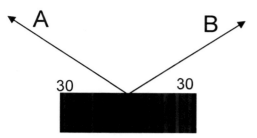

Problem: A block is held in static equilibrium by two cables. Suppose the tension in cables A and B are measured to be 50 Newtons each. The angle formed by each cable with the horizontal is 30 degrees. Calculate the weight of the block.

Solution: We know that the upward pull of the cable must balance the downward force of the weight of the block and the right pulling forces must balance the left pulling forces.

Using trigonometry we know that the y component of each cable can be calculated as:

$$F_y = 50 \text{ N sin } 30°$$
$$F_y = 25 \text{ N}$$

Since there are two cables supplying an upward force of 25 N each, the overall downward force supplied by the block must be 50 N.

Skill 9.2   Solve problems involving torque

Torque is rotational motion about an axis. It is defined as $\tau = L \times F$, where L is the lever arm. The length of the lever arm is calculated by measuring the perpendicular line drawn from the line of force to the axis of rotation.

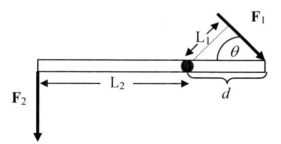

By convention, torques that act in a clockwise direction are considered negative and those in a counterclockwise direction are considered positive. In order for an object to be in equilibrium, the sum of the torques acting on it must be zero.

The equation that would put the above figure in equilibrium is $F_1 L_1 = -F_2 L_2$ (please note that in this case $L_1 = d\sin\theta$).

Examples:

Some children are playing with the spinner below, when one young boy decides to pull on the spinner arrow in the direction indicated by F_1. How much torque does he apply to the spinner arrow?

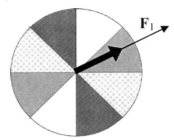

$\tau = L \times F$, but L=0 directly through the perpendicular the pivot point is 0). torque to the

because the line of force goes axis of rotation (i.e. the distance from the line of force to Therefore the boy applies no spinner.

Problem: In the system diagrammed below, find out what the magnitude of F_1 must be in order to keep the system in equilibrium.

Solution: For an object to be in equilibrium the forces acting on it must be balanced. This applies to linear as well as rotational forces known as moments or torques.

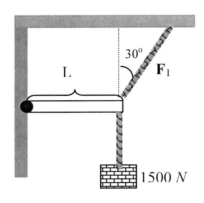

$$F_1 L = -F_2 L$$

$$F_{1x}L + F_{1y}L = -(-1500L)\ldots\text{but } F_{1x} = 0 \, *$$

$$F_{1y}L = 1500L$$

$$F_1 \cos 30 L = 1500L$$

$$(0.866)F_1 = 1500$$

$$F_1 = 1732N$$

the effect of $F_{1x}=0$ because that portion of the force goes right through the pivot and causes no torque.

Things to remember:
1. When writing the equation for a body at equilibrium, the point chosen for the axis of rotation is arbitrary.
2. The center of gravity is the point in an object where its weight can be considered to act for the purpose of calculating torque.
3. The lever arm, or moment arm, of a force is calculated as the **perpendicular** distance from the line of force to the pivot point/axis of rotation.
4. Counterclockwise torques are considered positive while clockwise torques are considered negative.

COMPETENCY 10.0 UNDERSTAND THE LAWS OF MOTION (INCLUDING RELATIVITY) AND CONSERVATION OF MOMENTUM

Skill 10.1 Explain the characteristics and give examples of each of Newton's laws of motion

Newton's first law of motion: "An object at rest tends to stay at rest and an object in motion tends to stay in motion with the same speed and in the same direction unless acted upon by an unbalanced force". Prior to Newton's formulation of this law, being at rest was considered the natural state of all objects, because at the earth's surface we have the force of gravity working at all times which causes nearly any object put into motion to eventually come to rest. Newton's brilliant leap was to recognize that an unbalanced force changes the motion of a body, whether that body begins at rest or at some non-zero speed.

We experience the consequences of this law everyday. For instance, the first law is why seat belts are necessary to prevent injuries. When a car stops suddenly, say by hitting a road barrier, the driver continues on forward until acted upon by a force. The seat belt provides that force and distributes the load across the whole body rather than allowing the driver to fly forward and experience the force against the steering wheel.

Example: A skateboarder is riding her skateboard down a road. The skateboard has a constant speed of 5 m/s. Then the skateboard hits a rock and stops suddenly. Since the rider has nothing to stop her when the skateboard stops, she

will continue to travel at 5 m/s until she hits the ground.

Newton's second law of motion: "The acceleration of an object as produced by a net force is directly proportional to the magnitude of the net force, in the same direction as the net force, and inversely proportional to the mass of the object". In the equation form, it is stated as $F = ma$, force equals mass times acceleration. It is important to remember that this is the net force and that forces are vector quantities. Thus if an object is acted upon by 12 forces that sum to zero, there is no acceleration. Also, this law embodies the idea of inertia as a consequence of mass. For a given force, the resulting acceleration is proportionally smaller for a more massive object because the larger object has more inertia.

Example:

A ball is dropped from a building. The mass of the ball is 2 kg. The acceleration of the object is 9.8 m/s^2 (gravitational acceleration). Therefore, the force acting on the ball is
$F = ma \Rightarrow F = 2$ kg x 9.8 m/s$^2 \Rightarrow F = 19.6$ N

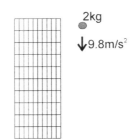

Newton's third law of motion: "For every action, there is an equal and opposite reaction". This statement means that in every interaction, there is a pair of forces acting on the two interacting objects. The size of the force on the first object equals the size of the force on the second object. The direction of the force on the first object is opposite to the direction of the force on the second object.

Example: A box is sitting on a table.

The mass of the box is 4 kg. Because of the effects of gravity, the box is applying a force of 39.2 N on the table. The table does not break or shift under the force of the box. This implies that the table is applying a force of 39.2 N on the box. Note that the force that the table is applying to the box is in the opposite direction to the force that the box is applying to the table.

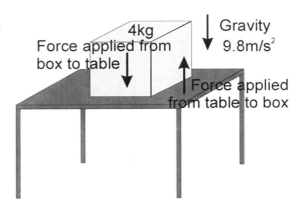

Here are a few more examples:
1. The propulsion/movement of fish through water: A fish uses its fins to push water backwards. The water pushes back on the fish. Because the force on the fish is unbalanced the fish moves forward.
2. The motion of car: A car's wheels push against the road and the road pushes back. Since the force of the road on the car is unbalanced the car moves forward.
3. Walking: When one pushes backwards on the foot with the muscles of the leg, the floor pushes back on the foot. If the forces of the leg on the foot and the floor on the foot are balanced, the foot will not move and the muscles of the body can move the other leg forward.

Skill 10.2   Apply Newton's laws of motion and the conservation of momentum in solving problems

Newton's laws of motion can be used together or separately to analyze a variety of physical situations. Simple examples are provided below and the importance of each law is highlighted.

Problem:
A 10 kg object moves across a frictionless surface at a constant velocity of 5 m/s. How much force is necessary to maintain this speed?

Solution:
Both Newton's first and second laws can help us understand this problem. First, the first law tells us that this object will continue its state of uniform speed in a straight line (since there is no force acting upon it). Additionally, the second law tells that because there is no acceleration (velocity is constant), no force is required. Thus, zero force is necessary to maintain the speed of 5 m/s.

Problem:
A car is driving down a road at a constant speed. The mass of the car is 400 kg. The force acting on the car is 4000 N and the force is in the same direction as the acceleration. What is the acceleration of the car?

Solution:

$$F = ma \implies a = \frac{F}{m} \implies a = \frac{4000\text{N}}{400kg} \implies a = 10m/s^2$$

Problem:
For the arrangement shown, find the force necessary to overcome the 500 N force pushing to the left and move the truck to the right with an acceleration of 5 m/s².

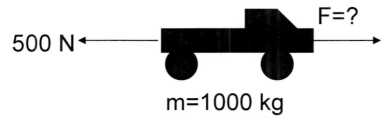

Solution:
The net force on the truck acting towards the right is F – 500N.
Using Newton's second law, F-500N = 1000kg x 5 m/s². Thus F = 5500N.

Problem:
An astronaut with a mass of 95 kg stands on a space station with a mass of 20,000 kg. If the astronaut is exerting 40 N of force on the space station, what is the acceleration of the space station and the astronaut?

Solution
To find the acceleration of the space station, we can simply apply Newton's second law:

$$A_s = \frac{F}{m_s} = \frac{40N}{20000kg} = 0.002 \, m/s^2$$

To find the acceleration of the astronaut, we must first apply Newton's third law to determine that the space station exerts an opposite force of -40 N on the astronaut. Here the minus sign simply denotes that the force is directed in the opposite direction. We can then calculate the acceleration, again using Newton's second law:

$$A_a = \frac{F}{m_a} = \frac{-40N}{95kg} = -0.42 \, m/s^2$$

The law of **conservation of momentum** states that the total momentum of an *isolated system* (not affected by external forces and not having internal dissipative forces) always remains the same. For instance, in any collision between two objects in an isolated system, the total momentum of the two objects after the collision will be the same as the total momentum of the two objects before the collision. In other words, any momentum lost by one of the objects is gained by the other. In one dimension this is easy to visualize.

Imagine two carts rolling towards each other as in the diagram below

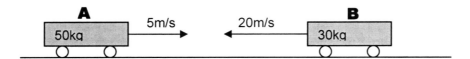

Before the collision, cart **A** has 250 kg m/s of momentum, and cart **B** has –600 kg m/s of momentum. In other words, the system has a total momentum of –350 kg m/s of momentum.

After the collision, the two cards stick to each other, and continue moving. How do we determine how fast, and in what direction, they go?

We know that the new mass of the cart is 80kg, and that the total momentum of the system is −350 kg m/s. Therefore, the velocity of the two carts stuck together must be $\frac{-350}{80} = -4.375\,m/s$

Conservation of momentum works the same way in two dimensions, the only change is that you need to use vector math to determine the total momentum and any changes, instead of simple addition.

Imagine a pool table like the one below. Both balls are 0.5 kg in mass.

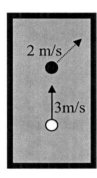

Before the collision, the white ball is moving with the velocity indicated by the solid line and the black ball is at rest.
After the collision the black ball is moving with the velocity indicated by the dashed line (a 135° angle from the direction of the white ball).

With what speed, and in what direction, is the white ball moving after the collision?

$p_{white\,/\,before} = .5 \cdot (0,3) = (0,1.5)$ $p_{black\,/\,before} = 0$ $p_{total\,/\,before} = (0,1.5)$

$p_{black\,/\,after} = .5 \cdot (2\cos 45, 2\sin 45) = (0.71, 0.71)$

$p_{white\,/\,after} = (-0.71, 0.79)$

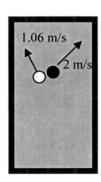

i.e. the white ball has a velocity of
$$v = \sqrt{(-.71)^2 + (0.79)^2} = 1.06\,m/s$$
and is moving at an angle of
$$\theta = \tan^{-1}\left(\frac{0.79}{-0.71}\right) = -48° \text{ from the horizontal}$$

Skill 10.3   Discuss the implications of special relativity for the laws of motion

Postulates of special relativity

Einstein's theory of special relativity built upon the foundation of Galilean relativity and incorporated an analysis of electromagnetics. Galilean relativity is based upon the concept that the laws of physics are invariant with respect to inertial frames of reference. In the context of classical electrodynamics, it is found that Maxwell's equations do not imply any variation in the speed of light with respect to the relative motion of the source and observer. As a result, the second fundamental postulate of special relativity is a statement of the invariance of the speed of light, c, with respect to inertial reference frames.

The postulates of special relativity in brief:
1. Special principle of relativity: The laws of physics are same in all inertial frames of reference.
2. Invariance of c: The speed of light in a vacuum is a universal constant for all observers, regardless of the inertial frame of reference or the relative motion of the source.

Force and acceleration

One of the implications of these postulates is that the relativistic mass m of a particle is dependent upon its velocity v. This relationship between the rest mass m_0 and the relativistic mass is expressed through the Lorentz factor γ.

$$m = \gamma m_0 = \frac{m_0}{\sqrt{1 - \dfrac{v^2}{c^2}}}$$

This has a particular effect on the relationship of force F and acceleration a, which is generally expressed using the momentum p.

$$F = \frac{\partial p}{\partial t} = \frac{\partial(mv)}{\partial t}$$

In the context of special relativity, where the mass is dependent on the velocity (which is a function of time), the above equation does not simplify to F = ma, as it does in Newtonian mechanics.

$$F = v\frac{\partial m}{\partial t} + ma = m_0\left(v\frac{\partial \gamma}{\partial t} + \gamma a\right)$$

Since the Lorentz factor increases asymptotically towards infinity as the speed of the particle or object approaches the speed of light, the force required to accelerate the particle also approaches infinity. It would then require infinite energy to accelerate an object to the speed of light, thus making c the "speed limit of the universe." In cases where the speed of the object is significantly less than c, the Lorentz factor is close to unity, thus making Newtonian mechanics an accurate approximation. (This concept is similar to the correspondence principle of quantum mechanics; in this case, relativistic mechanics becomes Newtonian mechanics in the limit as v goes to zero.)

Velocity
As mentioned, the postulates of special relativity seem to imply a universal speed limit. Thus, if this is to be the case in all inertial frames of reference, the observed speed of an object cannot be greater than the speed of light regardless of the velocities of the object and inertial frame of reference as measured in any other inertial frame of reference. As a result, although a particular frame of reference may be moving in one direction at speed w close to light speed and a particular object is moving in the opposite direction at a speed v, likewise near c, the speed of the object as measured from the reference frame *cannot* simply be the sum of v and w. Instead, special relativity uses a different approach to the calculation resulting in a relative velocity given by the following formula.

$$v' = \frac{v + w}{1 + vw/c^2}$$

Conservation of energy
In classical mechanics the behavior of particles or objects can be determined through the application of the conservation of energy and the conservation of momentum. Although, in this case, these two are seemingly independent concepts, in the case of relativistic mechanics they are shown to be interdependent. This interdependence results from the relationship of mass and energy that is derived from the application of the postulates of special relativity.

$$E^2 = \left(mc^2\right)^2 + \left(pc\right)^2$$

Thus, mass and energy (and, therefore, energy and momentum) are interdependent characteristics of the particle or object. The conservation laws for classical momentum and energy are then joined into a more general expression of the conservation of energy.

COMPETENCY 11.0 UNDERSTAND THE CHARACTERISTICS OF CIRCULAR MOTION AND SIMPLE HARMONIC MOTION, AND SOLVE PROBLEMS INVOLVING THESE TYPES OF MOTION

Skill 11.1 Applying vector analysis to describe uniform circular motion in radians

Speed remains constant in uniform circular motion. However, since the object is following a curve, the direction of the velocity is changing. Therefore, the object is accelerating. This acceleration is called **centripetal acceleration**.

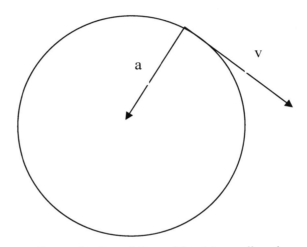

As seen in the above figure, the velocity of the object traveling in a circle is a tangent to the circle. The acceleration of the object is directed towards the center of the circle. Therefore, the object follows a curve.

In order to analyze a circle, it is helpful to use polar coordinates as reference.

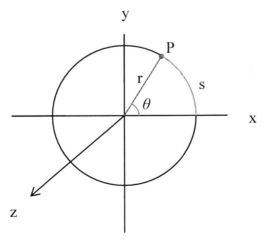

Using polar coordinates (r, θ), Point P on the circle can be defined by the distance from the origin r and the angle with respect to the x-axis θ.

The arc length s can be found using the following equation:

$$s = r\theta$$

where r is the radius and θ is the angular position.

The angular position θ is measure in radians. Note that if the angular position is given in degrees, they must be converted into radians first.

1 revolution = 360°= 2π radians

Angular velocity

Angular velocity is the angular displacement in a given time period. The instantaneous angular velocity can be found using the equation

$$\omega = \frac{d\theta}{dt}$$

where ω is the angular velocity, θ is the angular position and t is the time.

The relationship between linear velocity (v) and angular velocity (ω)

Imagine there are two revolving bodies that complete one revolution in the same amount of time. However the radius of the orbit of one of the bodies is twice that of the other. It is obvious that the object that has the larger orbit has a greater velocity as it needs to cover a larger distance in the same amount of time. Conversely, the angular velocity of the two bodies is the same as they sweep out equal angles in equal time. Therefore, linear velocity (v) is a function of the radius of a circle (r) for a constant angular velocity (ω).

If $s = r\theta$ then differentiating for time $\quad \dfrac{ds}{dt} = \dfrac{dr}{dt}\theta + \dfrac{d\theta}{dt}r$

This equation can be further simplified as the radius of the circle remains constant, therefore $\quad \dfrac{ds}{dt} = \dfrac{d\theta}{dt}r$

Since, $\omega = \dfrac{d\theta}{dt}$ and $\dfrac{ds}{dt} = v$, we have $v = \omega r$

Skill 11.2 Determining the magnitude and direction of the force acting on a particle in uniform circular motion

As explained in the previous section, in uniform circular motion the acceleration (a) is directed toward the center of the circular path and is always perpendicular to the velocity, as shown below:

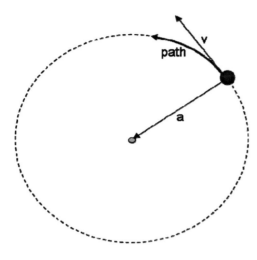

This centripetal acceleration is mathematically expressed as:

$$a = \frac{v^2}{r} = \frac{4\pi^2 r}{t^2}$$

where r is the radius of the circular path and t is the period of the motion or the time taken for the mass to travel once around the circle. The force (F) experienced by the mass (m) is known as centripetal force and is always directed towards the center of the circular path. It has constant magnitude given by the following equation:

$$F = ma = m\frac{v^2}{r}$$

Problem:
A car of mass 2000 Kg travels round a curve of radius 100m at a speed of 90 Km/h. What is the magnitude of the centripetal force acting on it?

Solution:

$$90 Km/h = \frac{90 \times 10^3}{60 \times 60} m/s = 25 m/s$$

$$a = \frac{v^2}{r} = \frac{(25)^2}{100} = 6.25 m/s^2$$

$$F = ma = 2000 \times 6.25 \ N = 12500 N$$

Skill 11.3   Analyzing the relationships among displacement, velocity, and acceleration in simple harmonic motion (e.g., simple pendulum, mass on a spring)

Simple harmonic (sinusoidal) motion involves a cyclical exchange of kinetic energy and potential energy as observed in a simple pendulum or a mass on a spring. The relationships among the various parameters of a system displaying simple harmonic motion depends on the type of system being examined. Once the displacement is known, the velocity and acceleration of the object undergoing harmonic motion can be calculated by calculating the first derivative with respect to time (for velocity) or the second derivative with respect to time (for acceleration).

The examples of the mass on a spring and the simple pendulum represent harmonic motion in one dimension and two dimensions, respectively. Each can be treated in a similar manner, although the details vary slightly. It can be shown, however, that the pendulum acts just like the mass on a spring when the displacement angle is small.

Linear oscillator (mass on a spring)
Given the frequency of oscillation f for a mass on a spring (a linear oscillator), along with the concomitant period T = 1/f, the displacement of the mass undergoing harmonic (sinusoidal) motion can be written as follows.

$$x(t) = x_{max} \cos(2\pi f t + \phi) = x_{max} \cos(\omega t + \phi)$$

Alternatively, $2\pi f$ can be written as the angular frequency ω. The coefficient x_{max} is the maximum displacement of the mass, and the term Φ is a phase constant that determines the position of the mass at time t = 0. If x(t) is differentiated once with respect to time, the velocity of the mass is revealed.

$$v(t) = \frac{\partial x(t)}{\partial t} = -\omega x_{max} \sin(\omega t + \phi)$$

Comparing the expressions for displacement and velocity, we can see that the velocity is maximized when the displacement is zero (all kinetic energy), and the displacement is maximized when the velocity is zero (all potential energy); that is, the displacement and velocity are 90^0 out of phase. The acceleration can be calculated by differentiating v(t) with respect to time.

$$a(t) = \frac{\partial v(t)}{\partial t} = -\omega^2 x_{max} \cos(\omega t + \phi) = -\omega^2 x(t)$$

The acceleration, as shown above, is in phase with the displacement. Applying Newton's second law of motion leads to Hooke's law, which relates the restoring force on the mass to the displacement x(t).

$$F = ma = -m\omega^2 x(t)$$

This equation may be expressed in terms of the so-called spring constant k, which is defined as $m\omega^2$.

$$F = ma = -k\, x(t)$$

Simple pendulum
The simple pendulum model provides a reasonably accurate representation of pendulum motion, especially in the case of small angular amplitude.

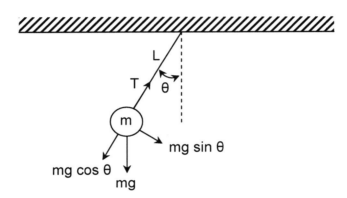

The restoring force F in pendulum motion is expressed as a component of the gravitational force mg perpendicular to the length of the string and is given by

$$F = -mg\sin\theta$$

The negative sign results from the force having a direction opposite to the displacement. The tension T on the string and the portion of the gravitational force in the opposite direction of T cancel one another.

In the case of small θ, sin θ is approximately equal to θ. The arc length traveled by the pendulum, s, is equal to the product of the length L of the string and the angle θ. Thus, the following expression can be derived.

$$F \approx -mg\theta = -mg\frac{s}{L} = -\frac{mg}{L}s$$

In the above equation, the force is shown to be of the same form as the linear harmonic oscillator, having, in this case, a "spring constant" of mg/L. As a result, the expressions found for the displacement, velocity and acceleration in the case of the linear oscillator can also be used here (in the case of small θ), where the frequency ω is replaced as follows.

$$\omega = \sqrt{\frac{g}{L}}$$

Skill 11.4   Solving problems involving springs and force constants

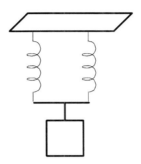

The above diagram is an example of a **Hookean system** (a spring, wire, rod, etc.) where the spring returns to its original configuration after being displaced and then released. When the spring is stretched a distance x, the restoring force exerted by the spring is expressed by Hooke's Law

$$F = -kx$$

The minus sign indicates that the restoring force is always opposite in direction to the displacement. The variable k is the spring constant and measures the stiffness of the spring in N/m.

The period of simple harmonic motion for a Hookean spring system is dependent upon the mass (m) of the spring and the stiffness of the spring (k) and is given by

$$T = \frac{1}{f} = \frac{1}{\frac{\omega}{2\pi}} = \frac{2\pi}{\omega} = \frac{2\pi}{\sqrt{\frac{k}{m}}} = 2\pi\sqrt{\frac{m}{k}}$$

Problem:
Each spring in the above diagram has a stiffness of $k = 20 N/m$. The mass of the object connected to the spring is 2 kg. Ignoring friction forces, find the period of motion.

Solution:
Utilizing Hooke's Law, the net restoring force on the spring would be

$$F = -(20N/m)x - (20N/m)x = -(40N/m)x$$

Comparison with $F = -kx$ shows the equivalent k to be 40 N/m.
Using the above formula for our problem, we have

$$T = 2\pi\sqrt{\frac{m}{k}} = 2\pi\sqrt{\frac{2kg}{40N/m}} = 2(3.14)\sqrt{0.05} = 1.4s$$

COMPETENCY 12.0   UNDERSTAND KEPLER'S LAWS AND THE LAW OF UNIVERSAL GRAVITATION, AND APPLY THEM TO SATELLITE MOTION

Skill 12.1   The geometric characteristics of planetary orbits

Johannes Kepler was a German mathematician who studied the astronomical observations made by Tyco Brahe. He derived the following three laws of planetary motion. Kepler's laws also predict the motion of comets.

Kepler's first law describes the shape of planetary orbits.   Specifically, the orbit of a planet is an ellipse that has the sun at one of the foci. Such an orbit looks like this:

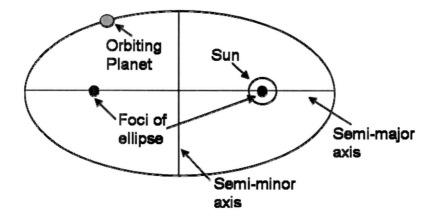

To analyze this situation mathematically, remember that the semi-major axis is denoted a, the semi-minor axis denoted b, and the general equation for an ellipse in polar coordinates is:

$$r = \frac{l}{1 + e\cos\theta}$$

Where r=radial coordinate
θ=angular coordinate
l= semi-latus rectum (l=b²/a)

e=eccentricity (for an ellipse, e=$\sqrt{1 - \dfrac{b^2}{a^2}}$

Thus, we can also determine the planet's maximum and minimum distance from the sun.

The point at which the planet is closest to the sun is known as the perihelion and occurs when θ=0:

$$r_{min} = \frac{l}{1+e}$$

The point at which the planet is farthest from the sun is known as the aphelion and occurs when θ=180°:

$$r_{max} = \frac{l}{1-e}$$

Skill 12.2 Apply Kepler's law of equal areas to solve problems involving satellite motion

Kepler's second law pertains to the relative speed of a planet as it orbits. This law says that a line joining the planet and the Sun sweeps out equal areas in equal intervals of time. In the diagram below, the two shaded areas demonstrate equal areas.

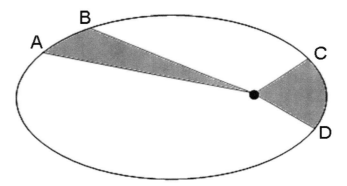

By Kepler's second law, we know that the planet will take the same amount of time to move between points A and B and between points C and D. Note that this means that the speed of the planet is inversely proportional to its distance from the sun (i.e., the plant moves fastest when it is closest to the sun). You can view an animation of this changing speed here:

http://home.cvc.org/science/kepler.gif

Kepler was only able to demonstrate the existence of this phenomenon but we now know that it is an effect of the Sun's gravity. The gravity of the Sun pulls the planet toward it thereby accelerating the planet as it nears. Using the first two laws together, Kepler was able to calculate a planet's position from the time elapsed since the perihelion.

Kepler's second law (the law of equal areas) can be used to predict the motion of satellites (including moons and planets). To begin, we must define several variables that describe the shape of the satellite's orbit. For variable definition, we will assume we are analyzing the motion of a planet about the sun and find the polar coordinates ((r,T); see below for definitions) of the planet as a function of time since perihelion.

s= the sun (one focus of the elliptical orbit)
z= the perihelion (the periapsis, the point at which the planet is closest the sun)
c= the center of the ellipse
p=the planet

Now define additional variables needed to analyze the orbit:

$a = |cz|$ → note that this is the semimajor axis of the ellipse

$\varepsilon = \dfrac{|cs|}{a}$ →this quantity is known as the eccentricity

$b = a\sqrt{1 - \varepsilon^2}$ →this is the semiminor axis of the ellipse
$r = |sp|$ →this quantity is the distance from the sun to the planet
$T = \angle zsp$ →this angle is known as the true anomaly

Then define x as the projection of the planet to an auxiliary circle:

$$|zsx| = \frac{a}{b} \cdot |zsp|$$

Also define y as a point on this auxiliary circle such that:

$$|zcy| = |zsx|$$

Then we can compute the mean anomaly, M:
$M = \angle zcy$

Finally, we can employ Kepler's assertion that the *area swept* since the perihelion is proportional to the *time* since the perihelion:

$$|zsp| = \frac{b}{a} \cdot |zsx| = \frac{b}{a} \cdot |zcy| = \frac{b}{a} \cdot |zcy| = \frac{b}{a} \cdot \frac{a^2 M}{2} = \frac{abM}{2}$$

$$M = \frac{2\pi t}{P}$$

where P=orbital period

Now we can compute the eccentric anomaly, E: $E = \angle zcx$

$$|zcy| = |zsx| = |zcx| - |scx|$$

$$\frac{a^2 M}{2} = \frac{a^2 E}{2} - \frac{a\varepsilon \cdot a \sin E}{2}$$

$$M = E - \varepsilon \cdot \sin E$$

This equation is known as Kepler's equation. However, because E cannot be isolated algebraically, it is typically solved using numerical methods.

Using trigonometric identities, we can next find an expression for the true anomaly, T (defined above):

$$\tan \frac{T}{2} = \sqrt{\frac{1+\varepsilon}{1-\varepsilon}} \cdot \tan \frac{E}{2}$$

Finally, we can compute the distance r:

$$r = a \cdot \frac{1 - \varepsilon^2}{1 + \varepsilon \cdot \cos T}$$

It is important to remember that Kepler's laws predict motion of the planets but they are simply more specific expressions of Newton's laws which predict the motion of all objects (on this scale). In fact, Kepler's laws can be derived from Newton's more general equations. To perform simple investigations of satellite motion, in which knowing the satellite's position and time at each point in its orbit is not necessary, it is often more straightforward to use Newton's laws.

This typically begins with the expression of net centripetal force:

$$F_{net} = \frac{M_{sat} v^2}{R}$$

which results from the gravity of the central body

$$F_{grav} = \frac{G \cdot M_{sat} \cdot M_{center}}{R^2}$$

$$\frac{M_{sat} \cdot v^2}{R} = \frac{G \cdot M_{sat} \cdot M_{center}}{R^2}$$

and so

$$v = \sqrt{\frac{G \cdot M_{center}}{R}} \qquad \frac{T^2}{R^3} = \frac{4 \cdot \pi^2}{G \cdot M_{center}}$$

Where F_{net}= net centripetal force
   M_{sat}=mass of the satellite
   M_{center}=mass of the body around which the satellite orbits
   G=gravitational constant (=6.67 x 10^{-11} N m^2/kg^2)
   R=average radius of the orbit
   v=velocity of the satellite
   T=period of the satellite's orbit

Example:
The Earth's moon returns to the same place in its orbital once every 27.2 days. What is its average speed and the average radius of its orbit? Use M_{Earth}=5.98x10^{24} kg and R_{Earth}=6.37x10^6 m.

Solution:
We have been provided that the period of the moon's orbit is 27.2 days, which is 2.35x10^6 seconds. Then we simply need to rearrange and solve the equations provided above.
To determine the radius:

$$\frac{T^2}{R^3} = \frac{4 \cdot \pi^2}{G \cdot M_{center}}$$

$$R^3 = \frac{T^2 \cdot G \cdot M_{center}}{4 \cdot \pi^2} = \frac{(2.35 \cdot 10^6\, s)^2 \cdot (6.67 \cdot 10^{-11}\, N \cdot m^2 / kg^2) \cdot (5.98 \cdot 10^{24}\, kg)}{4 \cdot \pi^2}$$

$$R^3 = 5.58 \cdot 10^{25}\, m^3$$

$$R = 3.82 \cdot 10^8\, m$$

Now for average speed:

$$v = \sqrt{\frac{G \cdot M_{center}}{R}} = \sqrt{\frac{(6.67 \cdot 10^{-11} \, N \cdot m^2 / kg^2) \cdot (5.98 \cdot 10^{24} \, kg)}{3.82 \cdot 10^8 \, m}}$$

$$v = 1.02 \cdot 10^3 \, m/s$$

Skill 12.3 **Apply Kepler's laws to relate the radius of a planet's orbit to its period of revolution**

Kepler's third law is also known as the harmonic law and it relates the size of a planet's orbital to the time needed to complete it. It states that the square of a planet's period is proportional to the cube of its mean distances from the Sun (this mean distance can be shown to be equal to the semi-major axis). So, we can state the third law as:

$$P^2 \propto a^3$$

where P=planet's orbital period (length of time needed to
complete one orbit)
a=semi-major axis of orbit

Furthermore, for two planets A and B:

$$P_A^2 / P_B^2 = a_A^3 / a_B^3$$

The units for period and semi-major axis have been defined such that $P^2 a^{-3}=1$ for all planets in our solar system. These units are sidereal years (yr) and astronomical units (AU). Sample values are given in the table below. Note that in each case $P^2 \sim a^3$

Planet	P (yr)	a (AU)	P^2	a^3
Venus	0.62	0.72	0.39	0.37
Earth	1.0	1.0	1.0	1.0
Jupiter	11.9	5.20	142	141

Skill 12.4 **Use the law of universal gravitation to interpret the relationship among force, mass, and the distance between masses**

See **Skill 8.3** for a discussion of Newton's law of universal gravitation.

COMPETENCY 13.0   APPLY THE PRINCIPLE OF CONSERVATION OF ENERGY AND THE CONCEPTS OF ENERGY, WORK, AND POWER

Skill 13.1   Analyze mechanical systems in terms of work, power, and conservation of energy

In physics, work is defined as force times distance $W = F \cdot s$. Work is a scalar quantity, it does not have direction, and it is usually measured in Joules ($N \cdot m$). It is important to remember, when doing calculations about work, that the only part of the force that contributes to the work is the part that acts in the direction of the displacement. Therefore, sometimes the equation is written as $W = F \cdot s \cos\theta$, where θ is the angle between the force and the displacement.

Problem:
A man uses 6N of force to pull a 10kg block, as shown below, over a distance of 3 m. How much work did he do?

Solution:

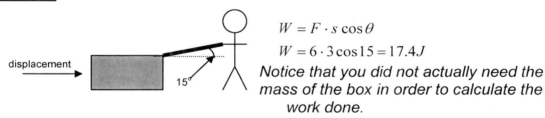

$$W = F \cdot s \cos\theta$$
$$W = 6 \cdot 3 \cos 15 = 17.4 J$$

Notice that you did not actually need the mass of the box in order to calculate the work done.

Energy is also defined, in relation to work, as the ability of an object to do work. As such, it is measured in the same units as work, usually Joules. Most problems relating work to energy are looking at two specific kinds of energy. The first, kinetic energy, is the energy of motion. The heavier an object is and the faster it is going, the more energy it has resulting in a greater capacity for work. The equation for kinetic energy is: $KE = \frac{1}{2}mv^2$.

Problem:
A 1500 kg car is moving at 60m/s down a highway when it crashes into a 3000kg truck. In the moment before impact, how much kinetic energy does the car have?

Solution:
$$KE = \frac{1}{2}mv^2 = \frac{1}{2} \cdot 1500 \cdot 60^2 = 2.7 \times 10^6 J$$

The other form of energy frequently discussed in relationship to work is gravitational potential energy, or potential energy, the energy of position. Potential energy is calculated as $PE = mgh$ where h is the distance the object is capable of falling.

Problem:
Which has more potential energy, a 2 kg box held 5 m above the ground or a 10 kg box held 1 m above the ground?

Solution:

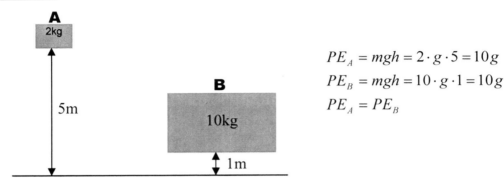

$$PE_A = mgh = 2 \cdot g \cdot 5 = 10g$$
$$PE_B = mgh = 10 \cdot g \cdot 1 = 10g$$
$$PE_A = PE_B$$

The power expended by a system can be defined as either the rate at which work is done by the system or the rate at which energy is transferred from it. There are many different measurements for power, but the one most commonly seen in physics problems is the Watt which is measured in Joules per second. Another commonly discussed unit of power is horsepower, and 1hp=746 W.

The **average power** of a system is defined as the work done by the system divided by the total change in time:

$\overline{P} = \dfrac{W}{\Delta t} \Rightarrow$ Where $\overline{P}$ = average power, W = work and Δt = change in time

The average power can also be written in terms of energy transfer $\Rightarrow \overline{P} = \dfrac{\Delta E}{\Delta t}$

and used the same way that the equation for work is used.

Problem:
A woman standing in her 4th story apartment raises a 10kg box of groceries from the ground using a rope. She is pulling at a constant rate, and it takes her 5 seconds to raise the box one meter. How much power is she using to raise the box?

Solution:

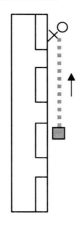

$$P = W/t$$

$$P = \frac{F \cdot s}{t} = \frac{mgh}{t} = \frac{10*9.8*1}{5} = 19.6W$$

Notice that because she is pulling at a constant rate, you don't need to know the actual distance she has raised the box. 2 meters in 10 seconds would give you the same result as 5 meters in 25 seconds.

Instantaneous power is the power measured or calculated in an instant of time. Since instantaneous power is the rate of work done when Δt is approaching 0s, the power is then written in derivate form:

$$P = \frac{dW}{dt} \Rightarrow$$ Where P = average power, dW = work and dt = change in time.

Since $W = Fs\cos\phi$, for a constant force the above equation can be written as:

$$P = \frac{dW}{dt} \Rightarrow P = \frac{d(Fs\cos\phi)}{dt} \Rightarrow P = \frac{(F\cos\phi)ds}{dt} \Rightarrow P = F\cos\phi\left(\frac{ds}{dt}\right) \Rightarrow P = Fv\cos\phi$$

where v is the velocity of the object.

Skill 13.2   Use the concept of conservation of energy to solve problems

According to the concept of conservation of energy, the energy in an isolated system remains the same although it may change in form. For instance, potential energy can become kinetic energy and kinetic energy, depending on the system, can become thermal or heat energy. Solving energy conservation problems depends on knowing the types of energy one is dealing with in a particular situation and assuming that the sum of all the different types of energy remains constant. Below we will discuss several different examples.

Example:
A rollercoaster at the top of a hill has a certain potential energy that will allow it to travel down the track at a speed based on its potential energy and friction with the track itself. At the bottom of the hill, when it has reached a stop, its potential energy is zero and all of the energy has been transferred from potential energy to kinetic energy (movement) and thermal energy (heat derived from friction). The equation below describes the relationship between potential energy and other forms of energy in this case:

Potential Energy = Kinetic energy(movement) + heat energy(friction)

Problem:
A skier travels down a ski slope with negligible friction. He begins at 100 meters in height, drops to a much lower level and ends at 90 meters in height. What is the skier's velocity at the 90 meter height?

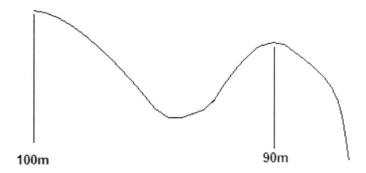

100m   90m

The skier's initial energy is only potential energy and is given by mgh = 100mg where m is the mass of the skier.
The skier's final energy is the sum of his potential and kinetic energies and is given by $mgh + 1/2mv^2 = 90mg + 1/2mv^2$ where v is the skier's velocity.
Using the principle of conservation of energy we know that the initial and final energies must be equal. Hence
$$100mg = 90mg + 1/2mv^2$$

Thus $1/2mv^2 = 10mg$; $1/2v^2 = 10g$; $v = 14m/s$ since $(g=9.81m/s^2)$

Problem:
d. A pebble weighing 10 grams is placed in a massless frictionless sling shot, with a spring constant of 200N/m, that is stretched back 0.5 meters. What is the total energy of the system before the pebble is released? What is the final height of the pebble if it is shot straight up and the effects of air resistance are negligible?

Solution:
If the initial height of the pebble is h = 0, total energy is given by
$E = \frac{1}{2} mv^2 + mgh + \frac{1}{2} kx^2 = 0 + 0 + 0.5(200)(0.5)^2 = 25$ Joules

At its final height, the velocity of the pebble will be zero. Since
$E = \frac{1}{2} mv^2 + mgh + \frac{1}{2} kx^2$, from the principle of conservation of energy
25 Joules = 0 + 0.010kg (9.81)h + 0 and h = 255 m

Skill 13.3   Determine power, mechanical advantage, and efficiency as they relate to work and energy in operations such as simple machines

Mechanical advantage is the multiplication of a force by a certain mechanism such as a lever or a pulley. Using such a device, a smaller force may be used to lift or move a heavier object.

The mechanical advantage of a lever is the ratio of the distance from the applied force (effort) to the fulcrum to the distance from the resistance force (load) to the fulcrum.

Example:

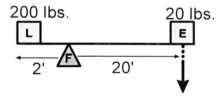

In the above diagram, the mechanical advantage would be $\frac{20}{2}$ or 10:1.

Therefore, an applied force of 20 lbs. will balance a resistance force of 200 lbs but only because it is applied at a distance 10 times greater from the fulcrum than the load.

The mechanical advantage of a pulley is equal to the number of ropes that support the pulley with each end of the rope counted as a separate rope.
Example:

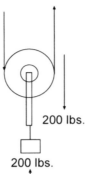

200 lbs.

200 lbs.

In the above diagram, there are two rope ends supporting the pulley. Therefore, the mechanical advantage of the pulley is 2 and an effort force of 100 lbs will lift a load of 200 lbs.

The mechanical advantage of an inclined plane is equal to the length of the slope divided by the height of the inclined plane. (This is true as long as the effort is applied parallel to the slope of the plane.)
Example:

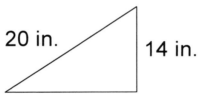

20 in.   14 in.

For the inclined plane in the above diagram, the mechanical advantage would be

$$\frac{S}{H} = \frac{20}{14} = 1.43$$

An inclined plane produces a mechanical advantage by increasing the distance through which the force must move. In this example, the object moved 20 in. along the slope in order to increase the vertical distance by 14 in.

Simple machines allow a user to apply less force to accomplish a task. The **work** done or **power** input, however, remains the same since the smaller force is typically applied over a greater distance. This satisfies the principle of conservation of energy.

The mechanical advantage discussed above is the theoretical mechanical advantage in the absence of dissipative forces such as friction. In the real world, friction eats up some of the input energy and results in a mechanical advantage that is smaller. The **efficiency** of a simple machine can be defined as the ratio of the actual mechanical advantage to the theoretical value expressed as a percentage. It is always less than 100%.

COMPETENCY 14.0 **UNDERSTAND THE DYNAMICS OF ROTATIONAL MOTION, INCLUDING TORQUE, ANGULAR MOMENTUM, MOTION WITH CONSTANT ANGULAR ACCELERATION, ROTATIONAL KINETIC ENERGY, CENTER OF MASS, AND MOMENT OF INERTIA**

Skill 14.1 **Explain the principles of motion with constant angular acceleration**

Linear motion is measured in rectangular coordinates. Rotational motion is measured differently, in terms of the angle of displacement. There are three common ways to measure rotational displacement; degrees, revolutions, and radians. Degrees and revolutions have an easy to understand relationship, one revolution is 360°. Radians are slightly less well known and are defined as $\frac{arc\ length}{radius}$. Therefore 360°=2π radians and 1 radian = 57.3°.

The major concepts of linear motion are duplicated in rotational motion with linear displacement replaced by **angular displacement** θ.

Angular velocity ω = rate of change of angular displacement.
Angular acceleration α = rate of change of angular velocity.

The kinematic equations for circular motion with constant angular acceleration are exactly analogous to the linear equations and are given by

$$\omega_f = \omega_i + \alpha t$$

$$\theta = \omega_i t + \frac{1}{2}\alpha t^2$$

$$\omega_f^2 = \omega_i^2 + 2\alpha\theta$$

Problem:
A wheel is rotating at the rate of 1 revolution in 8 seconds. If a constant deceleration is applied to the wheel it stops in 7 seconds. What is the deceleration applied to the wheel?

Solution:
Initial angular velocity = 2π/8 radians/sec = 0.25π radians/sec.
Final angular velocity = 0 radians/sec.
Using the rotational kinematic equations, angular acceleration applied to the wheel = (0 - 0.25x3.14)/7 = -0.11 radian/ (sec. sec)

One important difference in the angular equations relates to the use of mass in rotational systems. In rotational problems, not only is the mass of an object important but also its location. In order to include the spatial distribution of the mass of the object, a term called **moment of inertia** is used, $I = m_1 r_1^2 + m_2 r_2^2 + \cdots + m_n r_n^2$. The moment of inertia is always defined with respect to a particular axis of rotation.

Example:

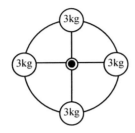

If the radius of the wheel on the left is 0.75m, what is its moment of inertia about an axis running through its center perpendicular to the plane of the wheel?

$$I = 3 \cdot 0.75^2 + 3 \cdot 0.75^2 + 3 \cdot 0.75^2 + 3 \cdot 0.75^2 = 6.75$$

Note: $I_{Sphere} = \frac{2}{5} mr^2$, $I_{Hoop/Ring} = mr^2$, $I_{disk} = \frac{1}{2} mr^2$

The relationship between angular acceleration α, moment of inertia I, and torque τ in any situation is given by the rotational analog of Newton's second law of motion:

$$\tau = I\alpha$$

Problem:
The angular velocity of a disc changes steadily from 0 to 6 rad/s in 200 ms. If the moment of inertia of the disc is 10 Kgm^2 what is the torque applied to the disc?

Solution:
The constant angular acceleration of the disc = $\dfrac{6}{200 \times 10^{-3}} = 30 rad / s^2$

Thus torque applied = $10 \times 30 = 300 N.m$.

<u>Problem:</u> A disk of mass 2 Kg and Radius 0.5m is mounted on a horizontal frictionless axle. A block of mass 1.5 Kg is hung from a massless string wrapped around the disk that does not slip. Find the acceleration of the block, the tension in the string and the angular acceleration of the disk.

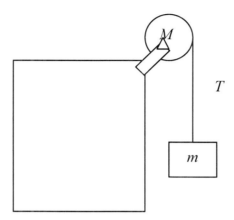

<u>Solution:</u> The forces on the block are the tension T in the string and the weight of the block. The acceleration a of the block is given by Newton's second law:

$$mg - T = ma$$

The torque on the disk of radius R is TR and the moment of inertia of the disk is $\frac{1}{2} MR^2$.

Thus, applying Newton's second law in angular form we get

$$TR = \frac{1}{2} MR^2 \left(\frac{a}{R} \right)$$

since the string does not slip and the linear acceleration of the disk is the same as the acceleration of the block. Combining both equations,

$$a = \frac{2T}{M} = \frac{2(mg - ma)}{M}; \Rightarrow a(M + 2m) = 2mg; \Rightarrow a = \frac{2mg}{M + 2m} = \frac{2 \times 1.5}{2 + 1.5 \times 2} \times 9.8 = 5.9 m/s^2$$

Using the second equation, the tension in the string is given by

$$T = \frac{1}{2} Ma = 0.5 \times 2 \times 5.9 = 5.9 N$$

The angular acceleration of the disk is given by

$$\alpha = \frac{a}{R} = \frac{5.9}{0.5} = 11.8 rad/s^2$$

Skill 14.2   The law of conservation of angular momentum

Angular momentum (L), and rotational kinetic energy (KE_r), are defined as follows: $L = I\omega, \quad KE_r = \dfrac{1}{2}I\omega^2$

According to the law of conservation of angular momentum, unless a net torque acts on a system, the angular momentum remains constant in both magnitude and direction. This can be used to solve many different types of problems including ones involving satellite motion.

Example:
A planet of mass m is circling a star in an orbit like the one below. If its velocity at point A is 60,000m/s, and $r_B = 8\ r_A$, what is its velocity at point B?

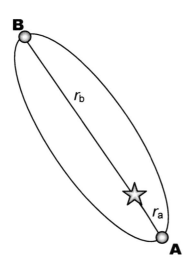

$$I_B\omega_B = I_A\omega_A$$

$$mr_B^2\omega_B = mr_A^2\omega_A$$

$$r_B^2\omega_B = r_A^2\omega_A$$

$$r_B^2\frac{v_B}{r_B} = r_A^2\frac{v_A}{r_A}$$

$$r_Bv_B = r_Av_A$$

$$8r_Av_B = r_Av_A$$

$$v_B = \frac{v_A}{8} = 7500m/s$$

Skill 14.3 Understand the concepts of center of mass, moment of inertia, and rotational kinetic energy

For a discussion of moment of inertia, see **Skill 14.1**. A concept related to the moment of inertia is the **radius of gyration** (*k*), which is the average distance of the mass of an object from its axis of rotation, i.e., the distance from the axis where a point mass *m* would have the same moment of inertia.

$$k_{Sphere} = \sqrt{\frac{2}{5}}r, \ k_{Hoop/Ring} = r, \ k_{disk} = \frac{r}{\sqrt{2}}.$$ As you can see $I = mk^2$

This is analogous to the concept of **center of mass**, the point where an equivalent mass of infinitely small size would be located, in the case of linear motion.

As with all systems, energy is conserved unless the system is acted on by an external force. This can be used to solve problems such as the one below.

Example:

A uniform ball of radius *r* and mass *m* starts from rest and rolls down a frictionless incline of height *h*. When the ball reaches the ground, how fast is it going?

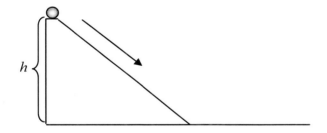

$$PE_{initial} + KE_{rotational/initial} + KE_{linear/initial} = PE_{final} + KE_{rotational/final} + KE_{linear/final}$$

$$mgh + 0 + 0 = 0 + \frac{1}{2}I\omega_{final}^2 + \frac{1}{2}mv_{final}^2 \rightarrow mgh = \frac{1}{2}\cdot\frac{2}{5}mr^2\omega_{final}^2 + \frac{1}{2}mv_{final}^2$$

$$mgh = \frac{1}{5}mr^2(\frac{v_{final}}{r})^2 + \frac{1}{2}mv_{final}^2 \rightarrow mgh = \frac{1}{5}mv_{final}^2 + \frac{1}{2}mv_{final}^2$$

$$gh = \frac{7}{10}v_{final}^2 \rightarrow v_{final} = \sqrt{\frac{10}{7}gh}$$

COMPETENCY 15.0 UNDERSTAND THE STATICS AND DYNAMICS OF FLUIDS

Skill 15.1 Applying the concepts of force, pressure, and density

Pressure is the force applied per unit area on a surface and is measured in the units N/m^2 in the SI system. Density is the mass per unit volume and is measured in the SI units kg/m^3. Pressure and density (in place of force and mass) are useful concepts when we are dealing with an extended mass of substance such as a fluid as opposed to a lump of matter such as a solid ball.

The weight of a column of fluid creates hydrostatic pressure. Common situations in which we might analyze hydrostatic pressure include tanks of fluid, a swimming pool, or the ocean. Also, **atmospheric pressure** is an example of hydrostatic pressure. Because hydrostatic pressure results from the force of gravity interacting with the mass of the fluid or gas, for an incompressible fluid it is governed by the following equation:

$$P = \rho g h$$

where P=hydrostatic pressure
ρ=density of the fluid
g=acceleration of gravity
h=height of the fluid column

Example: How much pressure is exerted by the water at the bottom of a 5 meter swimming pool filled with water?

Solution: We simply use the equation from above, recalling that the acceleration due to gravity is 9.8m/s² and the density of water is 1000 kg/m³.

$$P = \rho g h = 1000 \frac{kg}{m^3} \times 9.8 \frac{m}{s^2} \times 5m = 49,000\,Pa = 49kPa$$

According to **Pascal's principle**, when pressure is applied to an enclosed fluid, it is transmitted undiminished to all parts of the fluid. For instance, if an additional pressure P_0 is applied to the top surface of a column of liquid of height h as described above, the pressure at the bottom of the liquid will increase by P_0 and will be given by $P = P_0 + \rho g h$.

This principle is used in devices such as a **hydraulic lift** (shown below) which consists of two fluid-filled cylinders, one narrow and one wide, connected at the bottom. Pressure P (force = P X A1) applied on the surface of the fluid in the narrow cylinder is transmitted undiminished to the wider cylinder resulting in a larger net force (P X A2) transmitted through its surface. Thus a relatively small force is used to lift a heavy object. This does not violate the conservation of energy since the small force has to be applied through a large distance to move the heavy object a small distance.

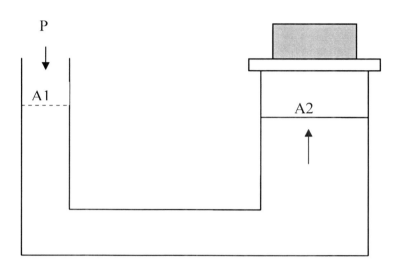

Skill 15.2   Use Bernoulli's principle to analyze fluid dynamics

The study of moving fluids is contained within fluid mechanics which is itself a component of continuum mechanics. Some of the most important applications of fluid mechanics involve liquids and gases moving in tubes and pipes.

Fluid flow may be laminar or turbulent. One cannot predict the exact path a fluid particle will follow in turbulent or erratic flow. Laminar flow, however, is smooth and each fluid particle follows a continuous path. Lines, know as **streamlines,** are drawn to show the path of a laminar fluid. Streamlines never cross one another and higher fluid velocity is depicted by drawing streamlines closer together.

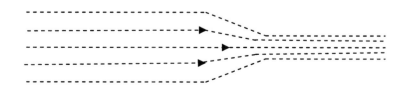

To understand the movement of laminar fluids, one of the first quantities we must define is volumetric flow rate which may have units of gallons per min (gpm), liters/s, cubic feet per min (cfm), gpf, or m^3/s:

$$Q = Av \cos \theta$$

Where Q=volumetric flow rate
A=cross sectional area of the pipe
v=fluid velocity
θ=the angle between the direction of the fluid flow and a vector normal to A

Note that in situations in which the fluid velocity is perpendicular to the cross sectional area, this equation is simply:

$$Q = Av$$

It is also convenient to sometimes discuss mass flow rate ($\dot{m}$), which we can easily find using the density (ρ) of the fluid:

$$\dot{m} = \rho v A = \rho Q$$

Usually, we make an assumption that the fluid is incompressible, that is, the density is constant. Like many commonly used simplifications, this assumption is largely and typically correct though real fluids are, of course, compressible to varying extents. When we do assume that density is constant, we can use conservation of mass to determine that when a pipe is expanded or restricted, the mass flow rate will remain the same. Let's see how this pertains to an example:

Given conservation of mass, it must be true that:

$$v_1 A_1 = v_2 A_2$$

This is known as the **equation of continuity**. Note that this means the fluid will flow faster in the narrower portions of the pipe and more slowly in the wider regions. An everyday example of this principle is seen when one holds their thumb over the nozzle of a garden hose; the cross sectional area is reduced and so the water flows more quickly.

Much of what we know about fluid flow today was originally discovered by Daniel Bernoulli. His most famous discovery is known as Bernoulli's Principle which states that, if no work is performed on a fluid or gas, an increase in velocity will be accompanied by a decrease in pressure. The mathematical statement of the Bernoulli's Principle for incompressible flow is:

$$\frac{v^2}{2} + gh + \frac{p}{\rho} = \text{constant}$$

where v= fluid velocity
g=acceleration due to gravity
h=height
p=pressure
ρ=fluid density

Though some physicists argue that it leads to the compromising of certain assumptions (i.e., incompressibility, no flow motivation, and a closed fluid loop), most agree it is correct to explain "lift" using Bernoulli's principle. This is because Bernoulli's principle can also be thought of as predicting that the pressure in moving fluid is less than the pressure in fluid at rest. Thus, there are many examples of physical phenomenon that can be explained by Bernoulli's Principle:

- The lift on airplane wings occurs because the top surface is curved while the bottom surface is straight. Air must therefore move at a higher velocity on the top of the wing and the resulting lower pressure on top accounts for lift.
- The tendency of windows to explode rather than implode in hurricanes is caused by the pressure drop that results from the high speed winds blowing across the outer surface of the window. The higher pressure on the inside of the window then pushes the glass outward, causing an explosion.
- The ballooning and fluttering of a tarp on the top of a semi-truck moving down the highway is caused by the flow of air across the top of the truck. The decrease in pressure causes the tarp to "puff up."
- A perfume atomizer pushes a stream of air across a pool of liquid. The drop in pressure caused by the moving air lifts a bit of the perfume and allows it to be dispensed.

Skill 15.3   Apply Archimedes' principle to solve problems involving buoyancy and flotation

Archimedes' Principle states that, for an object in a fluid, "the upthrust is equal to the weight of the displaced fluid" and the weight of displaced fluid is directly proportional to the volume of displaced fluid. The second part of his discovery is useful when we want to, for instance, determine the volume of an oddly shaped object. We determine its volume by immersing it in a graduated cylinder and measuring how much fluid is displaced. We explore his first observation in more depth below.

Today, we call Archimedes' "upthrust" **buoyancy**. Buoyancy is the force produced by a fluid on a fully or partially immersed object. The buoyant force (F$_{buoyant}$) is found using the following equation:

$$F_{bouyant} = \rho V g$$

> where ρ=density of the fluid
> V=volume of fluid displaced by the submerged object
> g=the acceleration of gravity

Notice that the buoyant force opposes the force of gravity. For an immersed object, a gravitational force (equal to the object's mass times the acceleration of gravity) pulls it downward, while a buoyant force (equal to the weight of the displaced fluid) pushes it upward.

Also note that, from this principle, we can predict *whether* an object will sink or float in a given liquid. We can simply compare the density of the material from which the object is made to that of the liquid. If the material has a lower density, it will float; if it has a higher density it will sink. Finally, if an object has a density equal to that of the liquid, it will neither sink nor float.

Example: Will gold (ρ=19.3 g/cm^3) float in water?

Solution: We must compare the density of gold with that of water, which is 1 g/cm^3.

$$\rho_{gold} > \rho_{water}$$

So, gold will sink in water.

Example: Imagine a 1 m^3 cube of oak (530 kg/m^3) floating in water. What is the buoyant force on the cube and how far up the sides of the cube will the water be?

Solution: Since the cube is floating, it has displaced enough water so that the buoyant force is equal to the force of gravity. Thus the buoyant force on the cube is equal to its weight 1X530X9.8 N = 5194 N.

To determine where the cube sits in the water, we simply the find the ratio of the wood's density to that of the water:

$$\frac{\rho_{oak}}{\rho_{water}} = \frac{530\,{kg}/{m^3}}{1000\,{kg}/{m^3}} = 0.53$$

Thus, 53% of the cube will be submerged. Since the edges of the cube must be 1m each, the top 0.47m of the cube will appear above the water.

COMPETENCY 16.0   **UNDERSTAND THE PRINCIPLES OF THE FIRST AND SECOND LAWS OF THERMODYNAMICS, THE RELATIONSHIP BETWEEN TEMPERATURE AND HEAT, AND THE PRINCIPLES OF THERMAL EXPANSION, THERMAL CONTRACTION, AND HEAT TRANSFER**

Skill 16.1   **Principles of the first and second laws of thermodynamics**

The first law of thermodynamics is a restatement of conservation of energy, i.e. the principle that energy cannot be created or destroyed. It also governs the behavior of a system and its surroundings. The change in heat energy supplied to a system (Q) is equal to the sum of the change in the internal energy (U) and the change in the work (W) done by the system against internal forces.

The internal energy of a material is the sum of the total kinetic energy of its molecules and the potential energy of interactions between those molecules. Total kinetic energy includes the contributions from translational motion and other components of motion such as rotation. The potential energy includes energy stored in the form of resisting intermolecular attractions between molecules.

Mathematically, we can express the relationship between the heat supplied to a system, its internal energy and work done by it as

$$\Delta Q = \Delta U + \Delta W$$

Let us examine a sample problem that relies upon this law.

A closed tank has a volume of 40.0 m^3 and is filled with air at 25°C and 100 kPa. We desire to maintain the temperature in the tank constant at 25°C as water is pumped into it. How much heat will have to be removed from the air in the tank to fill the tank ½ full?

Solution: The problem involves isothermal compression of a gas, so $\Delta U_{gas}=0$. Consulting the equation above, $\Delta Q = \Delta U + \Delta W$, it is clear that the heat removed from the gas must be equal to the work done by the gas.

$$Q_{gas} = W_{gas} = P_{gas}V_1 \ln\left(\frac{V_2}{V_T}\right) = P_{gas}V_T \ln\left(\frac{\tfrac{1}{2}V_T}{V_T}\right) = P_{gas}V_T \ln \tfrac{1}{2}$$

$$= (100kPa)(40.0m^3)(-0.69314) = -2772.58kJ$$

Thus, the gas in the tank must lose 2772.58 kJ to maintain its temperature.

To understand **the second law of thermodynamics**, we must first understand the concept of entropy. Entropy is the transformation of energy to a more disordered state and is the measure of how much energy or heat is available for work. The greater the entropy of a system, the less energy is available for work. The simplest statement of the second law of thermodynamics is that the entropy of an isolated system not in equilibrium tends to increase over time. The entropy approaches a maximum value at equilibrium. Below are several common examples in which we see the manifestation of the second law.

- The diffusion of molecules of perfume out of an open bottle
- Even the most carefully designed engine releases some heat and cannot convert all the chemical energy in the fuel into mechanical energy
- A block sliding on a rough surface slows down
- An ice cube sitting on a hot sidewalk melts into a little puddle; we must provide energy to a freezer to facilitate the creation of ice

When discussing the second law, scientists often refer to the "arrow of time". This is to help us conceptualize how the second law forces events to proceed in a certain direction. To understand the direction of the arrow of time, consider some of the examples above; we would never think of them as proceeding in reverse. That is, as time progresses, we would never see a puddle in the hot sun spontaneously freeze into an ice cube or the molecules of perfume dispersed in a room spontaneously re-concentrate themselves in the bottle. The above-mentioned examples are **spontaneous** as well as **irreversible**, both characteristic of increased entropy. . Entropy change is zero for a complete cycle in a **reversible process**, a process where infinitesimal quasi-static changes in the absence of dissipative forces can bring a system back to its original state without a net change to the system or its surroundings. All real processes are irreversible. The idea of a reversible process, however, is a useful abstraction that can be a good approximation in some cases.

Skill 16.2 Solve calorimetry problems involving heat capacity, specific heat, heat of fusion, and heat of vaporization

The **internal energy** of a material is the **sum of the total kinetic energy** of its molecules and the **potential energy** of interactions between those molecules. Total kinetic energy includes the contributions from translational motion and other components of motion such as rotation. The potential energy includes **energy stored in the form of resisting intermolecular attractions** between molecules. The **enthalpy** (*H*) of a material is the **sum of its internal energy and the mechanical work** it can do by driving a piston. A change in the **enthalpy** of a substance is the total **energy** change caused by **adding/removing heat** at constant pressure.

When a material is heated and experiences a phase change, **thermal energy is used to break the intermolecular bonds** holding the material together. Similarly, bonds are formed with the release of thermal energy when a material changes its phase during cooling. Therefore, **the energy of a material increases during a phase change that requires heat and decreases during a phase change that releases heat**. For example, the energy of H_2O increases when ice melts and decreases when water freezes.

Heat capacity and specific heat
A substance's molar **heat capacity** is the heat required to **change the temperature of one mole of the substance by one degree**. Heat capacity has units of joules per mol- kelvin or joules per mol- °C. The two units are interchangeable because we are only concerned with differences between one temperature and another. A Kelvin degree and a Celsius degree are the same size.

The **specific heat** of a substance (also called specific heat capacity) is the heat required to **change the temperature of one gram or kilogram by one degree**. Specific heat has units of joules per gram-°C or joules per kilogram-°C.

These terms are used to solve problems involving a change in temperature by applying the formula:

$q = n \times C \times \Delta T$ where $q \Rightarrow$ heat added (positive) or evolved (negative)

$\qquad n \Rightarrow$ amount of material

$\qquad C \Rightarrow$ molar heat capacity if *n* is in moles, specific heat if *n* is a mass

$\qquad \Delta T \Rightarrow$ change in temperature $T_{final} - T_{initial}$

Example:
What is the change in energy of 10 g of gold at 25 °C when it is heated beyond its melting point to 1300 °C. You will need the following data for gold:

Solid heat capacity: 28 J/mol-K

Molten heat capacity: 20 J/mol-K

Enthalpy of fusion: 12.6 kJ/mol

Melting point: 1064 °C

Solution: First determine the number of moles used: $10 \text{ g} \times \dfrac{1 \text{ mol}}{197 \text{ g}} = 0.051 \text{ mol}$.

There are then three steps. 1) Heat the solid. 2) Melt the solid. 3) Heat the liquid. All three require energy so they will be positive numbers.

1) Heat the solid:

$$q_1 = n \times C \times \Delta T = 0.051 \text{ mol} \times 28 \ \frac{J}{\text{mol-K}} \times (1064 \ °C - 25 \ °C)$$

$$= 1.48 \times 10^3 \text{ J} = 1.48 \text{ kJ}$$

2) Melt the solid: $q_2 = n \times \Delta H_{fusion} = 0.051 \text{ mol} \times 12.6 \ \dfrac{kJ}{mol}$

$$= 0.64 \text{ kJ}$$

3) Heat the liquid:

$$q_3 = n \times C \times \Delta T = 0.051 \text{ mol} \times 20 \frac{J}{\text{mol-K}} \times (1300 \ °C - 1064 \ °C)$$

$$= 2.4 \times 10^2 \text{ J} = 0.24 \text{ kJ}$$

The sum of the three processes is the total change in energy of the gold:

$$q = q_1 + q_2 + q_3 = 1.48 \text{ kJ} + 0.64 \text{ kJ} + 0.24 \text{ kJ} = 2.36 \text{ kJ}$$

$$= 2.4 \text{ kJ}$$

A **temperature vs. heat graph** can demonstrate these processes visually. One can also calculate the specific heat or latent heat of phase change for the material by studying the details of the graph.

Example: The plot below shows heat applied to 1g of ice at -40C. The horizontal parts of the graph show the phase changes where the material absorbs heat but stays at the same temperature. The graph shows that ice melts into water at 0C and the water undergoes a further phase change into steam at 100C.

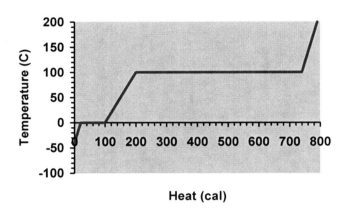

The specific heat of ice, water and steam and the latent heat of fusion and vaporization may be calculated from each of the five segments of the graph.

For instance, we see from the flat segment at temperature 0C that the ice absorbs 80 cal of heat. The latent heat L of a material is defined by the equation $\Delta Q = mL$ where ΔQ is the quantity of heat transferred and m is the mass of the material. Since the mass of the material in this example is 1g, the latent heat of fusion of ice is given by $L = \Delta Q / m = 80$ cal/g.

The next segment shows a rise in the temperature of water and may be used to calculate the specific heat C of water defined by $\Delta Q = mC\Delta T$, where ΔQ is the quantity of heat absorbed, m is the mass of the material and ΔT is the change in temperature. According to the graph, $\Delta Q = 200-100 = 100$ cal and $\Delta T = 100-0=100$C. Thus, $C = 100/100 = 1$ cal/gC.

Problem: The plot below shows the change in temperature when heat is transferred to 0.5g of a material. Find the initial specific heat of the material and the latent heat of phase change.

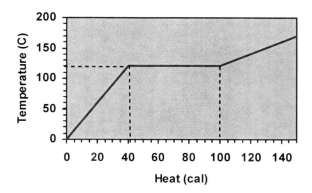

Solution: Looking at the first segment of the graph, we see that ΔQ = 40 cal and ΔT = 120 C. Since the mass m = 0.5g, the specific heat of the material is given by $C = \Delta Q / (m \Delta T)$ = 40/(0.5 X120) = 0.67 cal/gC. The flat segment of the graph represents the phase change. Here ΔQ = 100 - 40=60 cal. Thus, the latent heat of phase change is given by $L = \Delta Q / m$ = 60/(0.5) = 120 cal/g.

Skill 16.3   Analyze methods of heat transfer (i.e., conduction, convection, radiation) in practical situations

All heat transfer is the movement of thermal energy from hot to cold matter. This movement down a thermal gradient is a consequence of the second law of thermodynamics. The three methods of heat transfer are listed and explained below.

Conduction: Electron diffusion or photo vibration is responsible for this mode of heat transfer. The bodies of matter themselves do not move; the heat is transferred because adjacent atoms that vibrate against each other or as electrons flow between atoms. This type of heat transfer is most common when two solids come in direct contact with each other. This is because molecules in a solid are in close contact with one another and so the electrons can flow freely. It stands to reason, then, that metals are good conductors of thermal energy. This is because their metallic bonds allow the freest movement of electrons. Similarly, conduction is better in denser solids. Examples of conduction can be seen in the use of copper to quickly convey heat in cooking pots, the flow of heat from a hot water bottle to a person's body, or the cooling of a warm drink with ice.

The amount of heat transferred by conduction through a material depends on several factors. It is directly proportional to the temperature difference ΔT between the surface from which the heat is flowing and the surface to which it is transferred. Heat flow H increases with the area A through which the flow occurs and also with the time duration t. The thickness of the material reduces the flow of heat. The relationship between all these variables is expressed as

$$H = \frac{k.t.A.\Delta T}{d}$$

where the proportionality constant k is known as the **thermal conductivity**, a property of the material. Thermal conductivity of a good conductor is close to 1 (0.97 cal/cm.s.^0C for silver) while good insulators have thermal conductivity that is nearly zero (0.0005 cal/cm.s.^0C for wood).

Problem: A glass window pane is 50 cm long and 30 cm wide. The glass is 1 cm thick. If the temperature indoors is 15^0C higher than it is outside, how much heat will be lost through the window in 30 minutes? The thermal conductivity of glass is 0.0025 cal/cm.s.^0C.

Solution: The window has area A = 1500 sq. cm and thickness d = 1 cm. Duration of heat flow is 1800 s and the temperature difference $\Delta T = 15^0C$. Therefore heat loss through the window is given by

$$H = (0.0025 \times 1800 \times 1500 \times 15)/1 = 101250 \text{ calories}$$

Convection: Convection involves some conduction but is distinct in that it involves the movement of warm particles to cooler areas. Convection may be either natural or forced, depending on how the current of warm particles develops. Natural convection occurs when molecules near a heat source absorb thermal energy (typically via conduction), become less dense, and rise. Cooler molecules then take their place and a natural current is formed. Forced convection, as the name suggests, occurs when liquids or gases are moved by pumps, fans, or other means to be brought into contact with warmer or cooler masses. Because the free motion of particles with different thermal energy is key to this mode of heat transfer, convection is most common in liquid and gases. Convection can, however, transfer heat between a liquid or gas and a solid. Forced convection is used in "forced air" home heating systems and is common in industrial manufacturing processes. Additionally, natural convection is responsible for ocean currents and many atmospheric events. Finally, natural convection often arises in association with conduction, for instance in the air near a radiator or the water in a pot on the stove. The mathematical analysis of heat transfer by convection is far more complicated than for conduction or radiation and will not be addressed here.

Radiation: This method of heat transfer occurs via electromagnetic radiation. All matter warmer than absolute zero (that is, all known matter) radiates heat. This radiation occurs regardless of the presence of any medium. Thus, it occurs even in a vacuum. Since light and radiant heat are both part of the EM spectrum, we can easily visualize how heat is transferred via radiation. For instance, just like light, radiant heat is reflected by shiny materials and absorbed by dark materials. Common examples of radiant heat include the way sunlight travels from the sun to warm the earth, the use of radiators in homes, and the warmth of incandescent light bulbs.

The amount of energy radiated by a body at temperature T and having a surface area A is given by the Stefan-Boltzmann law expressed as

$$I = e\sigma AT^4$$

where I is the radiated power in watts, e (a number between 0 and 1) is the **emissivity** of the body and σ is a universal constant known as **Stefan's constant** that has a value of $5.6703 \times 10^{-8} W/m^2 . K^4$. Black objects absorb and radiate energy very well and have emissivity close to 1. Shiny objects that reflect energy are not good absorbers or radiators and have emissivity close to zero.

A body not only radiates thermal energy but also absorbs energy from its surroundings. The net power radiation from a body at temperature T in an environment at temperature T_0 is given by

$$I = e\sigma A(T^4 - T_0^4)$$

Problem: Calculate the net power radiated by a body of surface area 2 sq. m, temperature $30^0 C$ and emissivity 0.5 placed in a room at a temperature of $15^0 C$.

Solution: $I = 0.5 \times 5.67 \times 10^{-8} \times 2(303^4 - 288^4) = 88$ W

Skill 16.4   Solve problems involving thermal expansion and thermal contraction of solids

Most solid and liquid materials expand when heated with a change in dimension proportional to the change in temperature. A notable exception to this is water between 0^0C and 4^0C.

If we consider a long rod of length L that increases in length by ΔL when heated, the fractional change in length $\Delta L / L$ is directly proportional to the change in temperature ΔT.

$$\Delta L / L = \alpha . \Delta T$$

The constant of proportionality α is known as the **coefficient of linear expansion** and is a property of the material of which the rod is made.

Problem: The temperature of an iron rod 10 meters long changes from -3^0C to 12^0C. If iron has a coefficient of linear expansion of 0.000011 per ^0C, by how much does the rod expand?

Solution: The length of the rod L = 10 meters.
        Change in temperature $\Delta T = 12^0C - (-3^0C) = 15^0C$
        Change in length of the rod $\Delta L = 0.000011 \times 10 \times 15 = .00165$ meters

If instead of a rod, we consider an area A that increases by ΔA when heated, we find that the fractional change in area is proportional to the change in temperature ΔT. The proportionality constant in this case is known as the **coefficient of area expansion** and is related to the coefficient of linear expansion as demonstrated below.

If A is a rectangle with dimensions L_1 and L_2, then

$$A + \Delta A = (L_1 + \Delta L_1)(L_2 + \Delta L_2)$$
$$= (L_1 + \alpha L_1 \Delta T)(L_2 + \alpha L_2 \Delta T)$$
$$= L_1 L_2 + 2\alpha L_1 L_2 \Delta T + \alpha^2 (\Delta T)^2$$

Ignoring the higher order term for small changes in temperature, we find that

$$\Delta A = 2\alpha A \Delta T = \gamma A \Delta T$$

Thus the coefficient of area expansion $\gamma = 2\alpha$.

Following the same procedure as above, we can show that the change in volume of a material when heated may be expressed as

$$\Delta V = 3\alpha V \Delta T = \beta V \Delta T$$

where the **coefficient of volume expansion** $\beta = 3\alpha$.

Problem: An aluminum sphere of radius 10cm is heated from $0^0 C$ to $25^0 C$. What is the change in its volume? The coefficient of linear expansion of aluminum is 0.000024 per $^0 C$.

Solution: Volume V of the sphere = $\frac{4}{3}\Pi r^3 = \frac{4}{3} \times 3.14 \times 1000 cm^3 = 4186.67 cm^3$

Change in volume of the sphere = $3 \times 0.000024 \times 4186.67 \times 25 = 7.54 cm^3$

Skill 16.5   Use the principle of entropy to analyze the operation of heat engines (e.g., Carnot cycle)

A quantitative measure of entropy S is given by the statement that the change in entropy of a system that goes from one state to another in an isothermal and reversible process is the amount of heat absorbed in the process divided by the absolute temperature at which the process occurs.

$$\Delta S = \frac{\Delta Q}{T}$$

Stated more generally, the entropy change that occurs in a state change between two equilibrium states A and B via a reversible process is given by

$$\Delta S_{A \to B} = \int_{A}^{B} \frac{dQ}{T}$$

If we consider a **heat engine** that absorbs heat Q_h from a hot reservoir at temperature T_h and does work W while rejecting heat Q_c to a cold reservoir at a lower temperature T_c, $Q_h - Q_c = W$. The efficiency of the engine is the ratio of the work done to the heat absorbed and is given by

$$\varepsilon = \frac{W}{Q_h} = \frac{Q_h - Q_c}{Q_h} = 1 - \frac{Q_c}{Q_h}$$

It is impossible to build a heat engine with 100% efficiency, i.e. one where $Q_c = 0$.

Carnot described an ideal reversible engine, the **Carnot engine**, that works between two heat reservoirs in a cycle known as the **Carnot cycle** which consists of two isothermal (12 and 34) and two adiabatic processes (23 and 41) as shown in the diagram below.

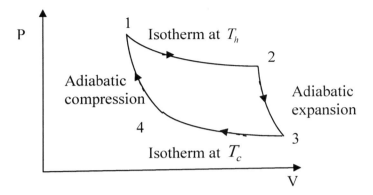

The efficiency of a Carnot engine is given by $\varepsilon = 1 - \dfrac{T_c}{T_h}$ where the temperature values are absolute temperatures. This is the highest efficiency that any engine working between T_c and T_h can reach. According to **Carnot's theorem**, no engine working between two heat reservoirs can be more efficient than a reversible engine. All such reversible engines have the same efficiency.

Combining the expression for efficiency of a Carnot engine (in terms of temperatures) with the expression for efficiency for a general heat engine (in terms of heat absorbed), we get:

$$1 - \frac{Q_c}{Q_h} = 1 - \frac{T_c}{T_h} \Rightarrow \frac{Q_c}{Q_h} = \frac{T_c}{T_h} \Rightarrow \frac{Q_c}{T_c} = \frac{Q_h}{T_h}$$

Thus the net change in entropy for the Carnot cycle is zero. This result can be generalized to any reversible cycle.

COMPETENCY 17.0 **UNDERSTAND THE KINETIC-MOLECULAR THEORY AND ITS RELATIONSHIP TO THERMODYNAMICS AND THE CHARACTERISTICS OF SOLIDS, LIQUIDS, AND GASES**

Skill 17.1 **Analyze the behavior of a gas in terms of the kinetic-molecular theory (i.e., ideal gas law)**

The relationship between **kinetic energy** and **intermolecular forces** determines whether a collection of molecules will be a gas, liquid, or solid. In a gas, the energy of intermolecular forces is much weaker than the kinetic energy of the molecules. Kinetic molecular theory is usually applied to gases and is best applied by imagining ourselves shrinking down to become a molecule and picturing what happens when we bump into other molecules and into container walls.

Gas **pressure** results from molecular collisions with container walls. The **number of molecules** striking an **area** on the walls and the **average kinetic energy** per molecule are the only factors that contribute to pressure. A higher **temperature** increases speed and kinetic energy. There are more collisions at higher temperatures, but the average distance between molecules does not change, and thus density does not change in a sealed container.

Kinetic molecular theory explains why the pressure and temperature of **ideal gases** behave the way they do by making a few assumptions, namely:

1) The energies of intermolecular attractive and repulsive forces may be neglected.
2) The average kinetic energy of the molecules is proportional to absolute temperature.
3) Energy can be transferred between molecules during collisions and the collisions are elastic, so the average kinetic energy of the molecules doesn't change due to collisions.
4) The volume of all molecules in a gas is negligible compared to the total volume of the container.

Strictly speaking, molecules also contain some kinetic energy by rotating or experiencing other motions. The motion of a molecule from one place to another is called **translation**. Translational kinetic energy is the form that is transferred by collisions, and kinetic molecular theory ignores other forms of kinetic energy because they are not proportional to temperature.

The following table summarizes the application of kinetic molecular theory to an increase in container volume, number of molecules, and temperature:

Effect of an **increase** in one variable with other two constant	Impact on gas: – = decrease, **0** = no change, **+** = increase						
	Average distance between molecules	Density in a sealed container	Average speed of molecules	Average translational kinetic energy of molecules	Collisions with container walls per second	Collisions per unit area of wall per second	Pressure (P)
Volume of container (V)	+	–	0	0	–	–	–
Number of molecules	–	+	0	0	+	+	+
Temperature (T)	0	0	+	+	+	+	+

Additional details on the kinetic molecular theory may be found at http://hyperphysics.phy-astr.gsu.edu/hbase/kinetic/ktcon.html. An animation of gas particles colliding is located at http://comp.uark.edu/~jgeabana/mol_dyn/.

The pressure, temperature and volume relationships for an ideal gas (a gas described by the assumptions of the kinetic molecular theory listed above) are given by the following gas laws:

Boyle's law states that the volume of a fixed amount of gas at constant temperature is inversely proportional to the gas pressure, or:

$$V \propto \frac{1}{P}.$$

Gay-Lussac's law states that the pressure of a fixed amount of gas in a fixed volume is proportional to absolute temperature, or:

$$P \propto T.$$

Charles's law states that the volume of a fixed amount of gas at constant pressure is directly proportional to absolute temperature, or:

$$V \propto T.$$

The **combined gas law** uses the above laws to determine a proportionality expression that is used for a constant quantity of gas:

$$V \propto \frac{T}{P}.$$

The combined gas law is often expressed as an equality between identical amounts of an ideal gas at two different states ($n_1 = n_2$):

$$\frac{P_1 V_1}{T_1} = \frac{P_2 V_2}{T_2}.$$

Avogadro's hypothesis states that equal volumes of different gases at the same temperature and pressure contain equal numbers of molecules. **Avogadro's law** states that the volume of a gas at constant temperature and pressure is directly proportional to the quantity of gas, or:

$$V \propto n \text{ where } n \text{ is the number of moles of gas.}$$

Avogadro's law and the combined gas law yield $V \propto \dfrac{nT}{P}$. The proportionality constant R--the **ideal gas constant**--is used to express this proportionality as the **ideal gas law**:

$$PV = nRT.$$

The ideal gas law is useful because it contains all the information of Charles's, Avogadro's, Boyle's, and the combined gas laws in a single expression.

Solving ideal gas law problems is a straightforward process of algebraic manipulation. **Errors commonly arise from using improper units**, particularly for the ideal gas constant R. An absolute temperature scale must be used—never °C—and is usually reported using the Kelvin scale, but volume and pressure units often vary from problem to problem.

If pressure is given in atmospheres and volume is given in liters, a value for R of **0.08206 L- atm/(mol- K)** is used. If pressure is given in pascal (newtons/m^2) and volume in m^3, then the SI value for R of **8.314 J/(mol- K)** may be used because a joule is defined as a newton- meter or a pascal- m^3. A value for R of **8.314 Pa- m^3/(mol- K)** is identical to the ideal gas constant using joules.

The ideal gas law may also be rearranged to determine gas molar density in moles per unit volume (molarity):

$$\frac{n}{V} = \frac{P}{RT}.$$

Gas density d in grams per unit volume is found after multiplication by the molecular weight M:

$$d = \frac{nM}{V} = \frac{PM}{RT}.$$

Molecular weight may also be determined from the density of an ideal gas:

$$M = \frac{dV}{n} = \frac{dRT}{P}.$$

Example: Determine the molecular weight of an ideal gas that has a density of 3.24 g/L at 800 K and 3.00 atm.

Solution:
$$M = \frac{dRT}{P} = \frac{\left(3.24 \ \frac{g}{L}\right)\left(0.08206 \ \frac{\text{L-atm}}{\text{mol-K}}\right)(800 \ \text{K})}{3.00 \ \text{atm}} = 70.9 \ \frac{g}{\text{mol}}.$$

Tutorials for gas laws may be found online at: http://www.chemistrycoach.com/tutorials-6.htm. A flash animation tutorial for problems involving a piston may be found at http://www.mhhe.com/physsci/chemistry/essentialchemistry/flash/gasesv6.swf.

Skill 17.2 Analyze phase changes in terms of kinetic-molecular theory and molecular structure

It is a common experience that increasing the temperature of an object, such as an ice cube or a piece of chocolate, leads to melting of the object. This is also known as a phase transition from solid to liquid. Such phenomena can be analyzed in terms of the kinetic-molecular theory and the molecular structure of the material. In this theory, the material is treated as a system of numerous particles whose collective composition and statistical motion can be used to determine the properties of the material.

According to the kinetic-molecular theory, the temperature of an object is actually a measurement of the average kinetic energy of the particles or molecules that compose it. As such, increasing or decreasing the kinetic energy leads to a corresponding increase or decrease in the temperature of the material. In a solid, which is at a relatively low temperature, the individual molecules have a low kinetic energy and, thus, relatively little motion. This allows for maintenance of the orderly, periodic arrangement of the molecules in the case of a crystal, for instance. As the temperature and, concomitantly, the kinetic energy are increased, the motion of the molecules becomes faster, and the bonds or arrangements that form the crystal (or other solid material) are increasingly broken. If the temperature becomes sufficiently high, the material can no longer be considered a solid, but has changed phase into either a liquid (through melting) or a gas (through sublimation). The types of bonds or arrangements in a particular material are determined by the characteristics of the atoms or molecules contained therein. Silicon, which is important for semiconductor applications, forms a diamond crystal lattice with covalent bonds. Other materials form different bonds or crystals depending on their molecular structures and characteristics.

The particular temperature at which phase changes take place is determined, in part, by the pressure on the material. Pressure and temperature, in the case of an ideal gas, are related in the kinetic-molecular theory by way of the ideal gas law, PV = nRT. **The relationships of the solid, liquid and gaseous phases of a material, relative to temperature and pressure, can be depicted by way of a phase diagram**. An example phase diagram is shown below.

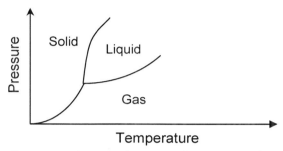

For the simple phase diagram above, the meeting point of the three curves is called the **triple point.** At this temperature and pressure, the phase of the material is not entirely distinguishable as either gas, liquid or solid. The particular details of the phase diagram are, again, determined by the molecular composition of the material and can generally be derived using the kinetic-molecular theory.

SUBAREA III   **ELECTRICITY AND MAGNETISM**

COMPETENCY 18.0   UNDERSTAND CHARACTERISTICS AND UNITS OF ELECTRIC CHARGE, ELECTRIC FIELDS, ELECTRIC POTENTIAL, AND CAPACITANCE; AND APPLY PRINCIPLES OF STATIC ELECTRICITY TO SOLVE PROBLEMS INVOLVING COULOMB'S LAW AND ELECTRIC FIELD INTENSITY

Skill 18.1   Analyze the behavior of an electroscope in given situations

An electroscope is, fundamentally, a charge-sensing device. Although it involves no digital output, it can be a helpful heuristic tool for understanding some basic concepts surrounding positive and negative charges. A simple design for an electroscope involves a glass container with a metal ball outside and a metal contact inside. Connected to the contact is a thin metal leaf that can be deflected by only a small applied force.

The behavior of the metal leaf inside the electroscope provides insight into the presence of charges in the surrounding environment or in the metal portions of the electroscope itself.

If a charged object is brought near (but does not touch) the electroscope (i.e., the metal ball), the free charge carriers in the metal rearrange themselves to minimize the energy of the system. If the object is negatively charged due to an excess of electrons, for example, the otherwise evenly distributed electrons in the metal of the electroscope will reorient such that a net positive charge is maintained near the charged object. This reorientation leads to a net negative charge in the other portion of the electroscope (i.e., inside the glass container), since the electroscope is initially uncharged. The local net charge results in electrical repulsion between the metal contact and the metal leaf. The leaf is then deflected.

If, while the charged object is near, the metal ball is touched by an observer, for example, the electroscope becomes "grounded," and some of the excess charge in certain areas is discharged. This results in the leaf returning to its original undeflected position. When the charged object is removed (after the observer also breaks contact), the result is that the metal leaf remains deflected. This phenomenon results from the excess charge obtained from grounding the electroscope during the time it was in the presence of a charged object. In the case of the above example, an excess of positive charge would remain and would disperse evenly throughout the metal (i.e., evenly very near the surface).

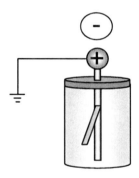 

If the electroscope is grounded once more, with the charged object no longer present, the excess charge will be lost and the leaf will return to its undeflected position.

In addition to the above method of charging the electroscope by induction, it may also be charged by rubbing the outside metal with a charged object, such as a glass rod. The conducting metal allows the excess charge on the rod to further distribute itself into a preferred lower-energy situation. This results in a deflection of the leaf that, as with the previous case, can be reversed by grounding the electroscope.

Skill 18.2 Apply Coulomb's law to determine the forces between charges

Any point charge may experience force resulting from attraction to or repulsion from another charged object. The easiest way to begin analyzing this phenomenon and calculating this force is by considering two point charges. Let us say that the charge on the first point is Q_1, the charge on the second point is Q_2, and the distance between them is r. Their interaction is governed by **Coulomb's Law** which gives the formula for the force F as:

$$F = k\frac{Q_1 Q_2}{r^2}$$

where $k = 9.0 \times 10^9 \frac{N \cdot m^2}{C^2}$ (known as Coulomb's constant)

The charge is a scalar quantity, however, the force has direction. For two point charges, the direction of the force is along a line joining the two charges. Note that the force will be repulsive if the two charges are both positive or both negative and attractive if one charge is positive and the other negative. Thus, a negative force indicates an attractive force.

When more than one point charge is exerting force on a point charge, we simply apply Coulomb's Law multiple times and then combine the forces as we would in any statics problem. Let's examine the process in the following example problem.

Problem: Three point charges are located at the vertices of a right triangle as shown below. Charges, angles, and distances are provided (drawing not to scale). Find the force exerted on the point charge A.

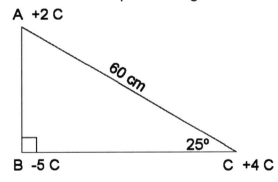

Solution: First we find the individual forces exerted on A by point B and point C. We have the information we need to find the magnitude of the force exerted on A by C.

$$F_{AC} = k\frac{Q_1 Q_2}{r^2} = 9 \times 10^9 \frac{N \cdot m^2}{C^2}\left(\frac{4C \times 2C}{(0.6m)^2}\right) = 2 \times 10^{11} N$$

To determine the magnitude of the force exerted on A by B, we must first determine the distance between them.

$$\sin 25° = \frac{r_{AB}}{60cm}$$
$$r_{AB} = 60cm \times \sin 25° = 25cm$$

Now we can determine the force.

$$F_{AB} = k\frac{Q_1 Q_2}{r^2} = 9 \times 10^9 \frac{N \cdot m^2}{C^2}\left(\frac{-5C \times 2C}{(0.25m)^2}\right) = -1.4 \times 10^{12} N$$

We can see that there is an attraction in the direction of B (negative force) and repulsion in the direction of C (positive force). To find the net force, we must consider the direction of these forces (along the line connecting any two point charges). We add them together using the law of cosines.

$$F_A^2 = F_{AB}^2 + F_{AC}^2 - 2F_{AB}F_{AC}\cos 75°$$
$$F_A^2 = (-1.4 \times 10^{12} N)^2 + (2 \times 10^{11} N)^2 - 2(-1.4 \times 10^{12} N)(2 \times 10^{11} N)^2 \cos 75°$$
$$F_A = 1.5 \times 10^{12} N$$

This gives us the magnitude of the net force, now we will find its direction using the law of sines.

$$\frac{\sin \theta}{F_{AC}} = \frac{\sin 75°}{F_A}$$
$$\sin \theta = F_{AC}\frac{\sin 75°}{F_A} = 2 \times 10^{11} N \frac{\sin 75°}{1.5 \times 10^{12} N}$$
$$\theta = 7.3°$$

Thus, the net force on A is 7.3° west of south and has magnitude 1.5 x 10^{12}N. Looking back at our diagram, this makes sense, because A should be attracted to B (pulled straight south) but the repulsion away from C "pushes" this force in a westward direction.

Skill 18.3 **Apply principles of electrostatics to determine electric field intensity**

An electric field is a vector field that is characterized both by a magnitude and a direction. The vector nature of the electric field arises from the fact that it is essentially the force that a unit positive charge would experience if placed in the field. Even though the field exists in the absence of any such actual charge, the magnitude and direction of the field are defined in this way.

Electric fields can be generated by a single point charge or by a collection of charges in close proximity. The electric field generated from a point charge is given by:

$$E = \frac{kQ}{r^2}$$

where E= the electric field

$$k = 9.0 \times 10^9 \, \frac{N \cdot m^2}{C^2} \text{ (Coulomb's constant)}$$

Q= the point charge
r= distance from the charge

Electric fields are visualized with field lines, which demonstrate the strength and direction of an electric field. The electric field around a positive charge points away from the charge and the electric field around a negative charge points toward the charge.

While it's easy enough to calculate and visualize the field generated by a single point charge, we can also determine the nature of an electric field produced by a collection of charge simply by adding the vectors from the individual charges. This is known as the superposition principle. The following equation demonstrates how this principle can be used to determine the field resulting from hundreds or thousands of charges.

$$\vec{E}_{\text{total}} = \sum_i \vec{E}_i = \vec{E}_1 + \vec{E}_2 + \vec{E}_3 \ldots$$

Skill 18.4   Explore the relationships between capacitance, charge, and potential difference

Capacitance (C) is a measure of the stored electric charge per unit electric potential. The mathematical definition is:

$$C = \frac{Q}{V}$$

It follows from the definition above that the units of capacitance are coulombs per volt, a unit known as a farad ($F=C/V$). In circuits, devices called parallel plate capacitors are formed by two closely spaced conductors. The function of capacitors is to store electrical energy. When a voltage is applied, electrical charges build up in both the conductors (typically referred to as plates). These charges on the two plates have equal magnitude but opposite sign. The capacitance of a capacitor is a function of the distance d between the two plates and the area A of the plates:

$$C \approx \frac{\varepsilon A}{d}; A \gg d^2$$

Capacitance also depends on the permittivity of the non-conducting matter between the plates of the capacitor. This matter may be only air or almost any other non-conducting material and is referred to as a **dielectric**. The permittivity of empty space ε_0 is roughly equivalent to that for air, $\varepsilon_{air}=8.854 \times 10^{-12}$ C ^{2}/N•m^2. For other materials, the dielectric constant, κ, is the permittivity of the material in relation to air ($\kappa=\varepsilon/\varepsilon_{air}$). The make-up of the dielectric is critical to the capacitor's function because it determines the maximum energy that can be stored by the capacitor. This is because an overly strong electric field will eventually destroy the dielectric.

In summary, a capacitor is "charged" as electrical energy is delivered to it and opposite charges accumulate on the two plates. The two plates generate electric fields and a voltage develops across the dielectric. The energy stored in the capacitor, then, is equal to the amount of work necessary to create this voltage. The mathematical statement of this is:

$$E_{stored} = \frac{1}{2}CV^2 = \frac{1}{2}\frac{Q^2}{C} = \frac{1}{2}VQ$$

The work per unit volume or the **electric field energy density** within a capacitor can be shown to be $\eta = \frac{1}{2}\varepsilon E^2$. This result is generally valid for the energy per unit volume of any electrostatic field, not only for a constant field within a capacitor.

Problem: Imagine that a parallel plate capacitor has an area of 10.00 cm^2 and a capacitance of 4.50 pF. The capacitor is connected to a 12.0 V battery. The capacitor is completely charged and then the battery is removed. What is the separation of the plates in the capacitor? How much energy is stored between the plates? We've assumed that this capacitor initially had no dielectric (i.e., only air between the plates) but now imagine it has a Mylar dielectric that fully fills the space. What will the new capacitance be? (for Mylar, □=3.5)

Solution: To determine the separation of the plates, we use our equation for a capacitor:

$$C = \frac{\varepsilon_0 A}{d}$$

We can simply solve for d and plug in our values:

$$d = \varepsilon_0 \frac{A}{C} = \left(8.854 \times 10^{-12} \frac{C}{N \cdot m^2}\right) \frac{10 \times 10^{-4} m^2}{4.5 \times 10^{-12} F} = 1.97 \times 10^{-3} m = 1.97 mm$$

Similarly, to find stored energy, we simply employ the equation above:

$$E_{stored} = \frac{1}{2} QV$$

But we don't yet know the charge Q, so we must first find it from the definition of capacitance:

$$C = \frac{Q}{V}$$

$$Q = CV = (4.5 \times 10^{-12}) \times (12V) = 5.4 \times 10^{-11} C$$

Now we can find the stored energy:

$$E_{stored} = \frac{1}{2} QV = \frac{1}{2} (5.4 \times 10^{-11} C)(12V) = 3.24 \times 10^{-10} J$$

To find the capacitance with a Mylar dielectric, we again use the equation for capacitance of a parallel plate capacitor. Note that the new capacitance can be found by multiplying the original capacitance by κ:

$$C = \frac{\kappa_{Mylar} \varepsilon_0 A}{d} = \kappa_{Mylar} C_0 = 3.5 \times 4.5 pF = 15.75 pF$$

PHYSICS   120

**COMPETENCY 19.0   UNDERSTAND CHARACTERISTICS OF ELECTRIC
CURRENT AND COMPONENTS OF ELECTRIC
CIRCUITS**

**Skill 19.1   Analyze a DC circuit in terms of conservation of energy and
conservation of charge (i.e., Kirchhoff's law, Ohm's law)**

Ohm's Law is the most important tool we posses to analyze electrical circuits.
Ohm's Law states that the current passing through a conductor is directly
proportional to the voltage drop and inversely proportional to the resistance of the
conductor. Stated mathematically, this is:

$$V = IR$$

Problem:
The circuit diagram at right shows
three resistors connected to a battery
in series. A current of 1.0A flows
through the circuit in the direction
shown. It is known that the equivalent
resistance of this circuit is 25 Ω. What
is the total voltage supplied by the
battery?

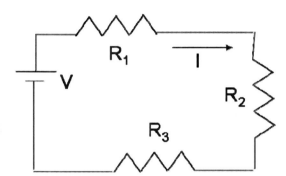

Solution:

To determine the battery's voltage, we simply apply Ohm's Law:

$$V = IR = 1.0A \times 25\Omega = 25V$$

Kirchoff's Laws are a pair of laws that apply to conservation of charge and
energy in circuits and were developed by Gustav Kirchoff.

Kirchoff's Current Law: At any point in a circuit where charge density is
constant, the sum of currents flowing toward the point must be equal to the sum
of currents flowing away from that point.

Kirchoff's Voltage Law: The sum of the electrical potential differences around a
circuit must be zero.

While these statements may seem rather simple, they can be very useful in
analyzing DC circuits, those involving constant circuit voltages and currents.

Problem:
The circuit diagram at right shows three resistors connected to a battery in series. A current of 1.0 A is generated by the battery. The potential drop across R_1, R_2, and R_3 are 5V, 6V, and 10V. What is the total voltage supplied by the battery?

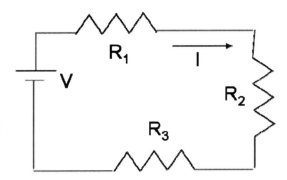

Solution:
Kirchoff's Voltage Law tells us that the total voltage supplied by the battery must be equal to the total voltage drop across the circuit. Therefore:

$$V_{battery} = V_{R_1} + V_{R_2} + V_{R_3} = 5V + 6V + 10V = 21V$$

Problem:
The circuit diagram at right shows three resistors wired in parallel with a 12V battery. The resistances of R_1, R_2, and R_3 are 4 Ω, 5 Ω, and 6 Ω, respectively. What is the total current?

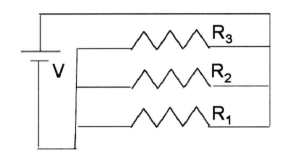

Solution:
This is a more complicated problem. Because the resistors are wired in parallel, we know that the voltage entering each resistor must be the same and equal to that supplied by the battery. We can combine this knowledge with **Ohm's Law** to determine the current across each resistor:

$$I_1 = \frac{V_1}{R_1} = \frac{12V}{4\Omega} = 3A$$

$$I_2 = \frac{V_2}{R_2} = \frac{12V}{5\Omega} = 2.4A$$

$$I_3 = \frac{V_3}{R_3} = \frac{12V}{6\Omega} = 2A$$

Finally, we use Kirchoff's Current Law to find the total current:

$$I = I_1 + I_2 + I_3 = 3A + 2.4A + 2A = 7.4A$$

Resistors and capacitors may also be combined together in circuits. Below we will consider the simplest RC circuit with a resistor and a capacitor in series connected to a voltage source.

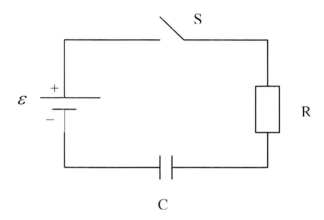

Applying Kirchhoff's first rule we get $\varepsilon = V_R + V_C = IR + \dfrac{Q}{C} = R\dfrac{dQ}{dt} + \dfrac{Q}{C}$ since the current I in the circuit is equal to the rate of increase of charge on the capacitor. If the charge on the capacitor is zero when the switch is closed at time t=0, the current I in the circuit starts out at the value ε/R and gradually falls to zero as the capacitor charge gradually rises to its maximum value. The mathematical expressions for these two quantities are

$$Q(t) = C\varepsilon(1 - e^{-t/RC}); I(t) = \frac{\varepsilon}{R}e^{-t/RC}$$

The product RC is known as the **time constant** of the circuit.

If the battery is now removed and the switch closed at t=0, the capacitor will slowly discharge with its charge at any point t given by

$Q(t) = Q_f e^{-t/RC}$ where Q_f is the charge that the capacitor started with at t=0.

Skill 19.2   Discuss factors that affect resistance

Conductors are those materials which allow for the free passage of electrical current. However, all materials exhibit a certain opposition to the movement of electrons. This opposition is known as resistivity (ρ). Resistivity is determined experimentally by measuring the resistance of a uniformly shaped sample of the material and applying the following equation:

$$\rho = R \frac{A}{l}$$

where ρ = static resistivity of the material
R = electrical resistance of the material sample
A = cross-sectional area of the material sample
L = length of the material sample

The temperature at which these measurements are taken is important as it has been shown that resistivity is a function of temperature. For conductors, resistivity increases with increasing temperature and decreases with decreasing temperature. At extremely low temperatures resistivity assumes a low and constant value known as residual resistivity (ρ_0). Residual resistivity is a function of the type and purity of the conductor.

The following equation allows us to calculate the resistivity ρ of a material at any temperature given the resistivity at a reference temperature, in this case at $20^0 C$:

$$\rho = \rho_{20}[1 + a(t - 20)]$$

where ρ_{20} = resistivity at $20^0 C$

a = proportionality constant characteristic of the material

t =temperature in Celsius

Problem: The tungsten filament in a certain light bulb is a wire 8 μm in diameter and 10 mm long. Given that, for tungsten, $\rho_{20} = 5.5 \times 10^{-8}$ Ω·m and $a = 4.5 \times 10^{-3} K^{-1}$, what will the resistance of the filament be at 45°C?

Solution: First we must find the resistivity of the tungsten at 45°C:

$$\rho = 5.5 \times 10^{-8}\left(1 + 4.5 \times 10^{-3}(45 - 20)\right) = 6.1 \times 10^{-8} \Omega.m$$

Now we can rearrange the equation defining resistivity and solve for the resistance of the filament:

$$R = \rho\frac{l}{A} = 6.1 \times 10^{-8} \times 0.01 / (\pi(4 \times 10^{-6})^2) = 12.1\Omega$$

Skill 19.3 Give examples of schematic diagrams of electric circuits

See **Skills 19.1** and **19.4** for examples of schematic diagrams of electric circuits.

Skill 19.4 Apply principles of DC circuits to reduce a complex circuit to a simpler equivalent circuit

Below is a diagram demonstrating a simple circuit with resistors in parallel (on right) and in series (on left). Note the symbols used for a battery (noted V) and the resistors (noted R).

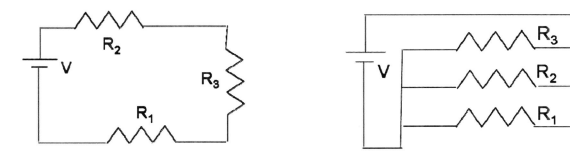

When the resistors are placed in series, the current through each one will be the same. When they are placed in parallel, the voltage through each one will be the same. To understand basic circuitry, it is important to master the rules by which the equivalent resistance (R_{eq}) or capacitance (C_{eq}) can be calculated from a number of resistors or capacitors. Several resistors in series or parallel can be replaced by one resistor with equivalent resistance, thus simplifying the circuit.

Resistors in parallel: $\dfrac{1}{R_{eq}} = \dfrac{1}{R_1} + \dfrac{1}{R_2} + \cdots + \dfrac{1}{R_n}$

Resistors in series: $R_{eq} = R_1 + R_2 + \cdots + R_n$

Capacitors in parallel: $C_{eq} = C_1 + C_2 + \cdots + C_n$

Capacitors in series: $\dfrac{1}{C_{eq}} = \dfrac{1}{C_1} + \dfrac{1}{C_2} + \cdots + \dfrac{1}{C_n}$

Example:
Three resistors are connected to a battery as shown. Find an expression for the current in the circuit.

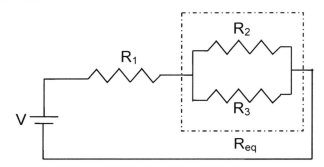

Solution:
Here we see that R_2 and R_3 are in parallel with one another and in series with R_1. Thus, we should begin by determining the equivalent resistance, R_{eq}, for R_2 and R_3 (illustrated with a dashed line). Because these resistors are in parallel:

$$\frac{1}{R_{eq}} = \frac{1}{R_2} + \frac{1}{R_3}$$

$$R_{eq} = \frac{R_2 \times R_3}{R_2 + R_3}$$

Now, we essentially have a situation in which R_{eq} is in series with R_1, so we can find the total resistance of the circuit as:

To find the current circuit, we simply

$$R_{total} = R_1 + \frac{R_2 \times R_3}{R_2 + R_3}$$

flowing through the apply Ohm's law:

$$I = \frac{V}{R_1 + \dfrac{R_2 \times R_3}{R_2 + R_3}}$$

Now let's analyze a more complicated mixed parallel-series circuit, this time involving capacitors.

Example:
Find the total equivalent capacitance of this circuit.

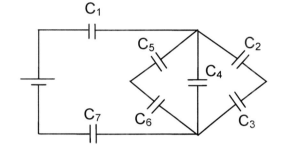

Solution:
This circuit is drawn in such a way that identifying the parallel and series relationships is difficult. The first step in simplifying it is to redraw it to elucidate those relationships.

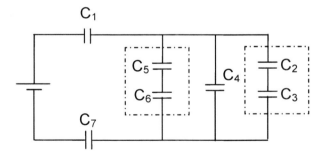

Now we can begin to simplify the circuit in a step-wise fashion. We begin by combining C_5 with C_6 and C_2 with C_3 (these are pairs in series).

$$C_{5,6} = \left(\frac{1}{C_5} + \frac{1}{C_6} \right)^{-1} \qquad C_{2,3} = \left(\frac{1}{C_2} + \frac{1}{C_3} \right)^{-1}$$

The next steps will involve combining $C_{5,6}$ and $C_{2,3}$ with C_4 in parallel and the resulting equivalent capacitance in series with C_1 and C_7.

COMPETENCY 20.0   **UNDERSTAND MAGNETS, ELECTROMAGNETS, AND MAGNETIC FIELDS; THE EFFECTS OF MAGNETIC FIELDS ON MOVING ELECTRIC CHARGES; AND THE APPLICATIONS OF ELECTROMAGNETISM**

Skill 20.1   Magnets and magnetism

Magnetism is a phenomenon in which certain materials, known as magnetic materials, attract or repel each other. A magnet has two poles, a south pole and a north pole. Like poles repel while unlike poles attract. Magnetic poles always occur in pairs known as **magnetic dipoles**. One cannot isolate a single magnetic pole. If a magnet is broken in half, opposite poles appear at both sides of the break point so that one now has two magnets each with a south pole and a north pole. No matter how small the pieces a magnet is broken into, the smallest unit at the atomic level is still a dipole.

A large magnet can be thought of as one with many small dipoles that are aligned in such a way that apart from the pole areas, the internal south and north poles cancel each other out. Destroying this long range order within a magnet by heating or hammering can demagnetize it. The dipoles in a **non-magnetic** material are randomly aligned while they are perfectly aligned in a preferred direction in **permanent** magnets. In a **ferromagnet**, there are domains where the magnetic dipoles are aligned, however, the domains themselves are randomly oriented. A ferromagnet can be magnetized by placing it in an external magnetic field that exerts a force to line up the domains. A magnet produces a magnetic field that exerts a force on any other magnet or current-carrying conductor placed in the field.

Skill 20.2 Determining the orientation and magnitude of a magnetic field

Magnetic field lines are a good way to visualize a magnetic field. The distance between magnetic fields lines indicates the strength of the magnetic field such that the lines are closer together near the poles of the magnets where the magnetic field is the strongest. The lines spread out above and below the middle of the magnet, as the field is weakest at those points furthest from the two poles. The SI unit for magnetic field known as magnetic induction is Tesla(T) given by 1T = 1 N.s/(C.m) = 1 N/(A.m). Magnetic fields are often expressed in the smaller unit Gauss (G) (1 T = 10,000 G). Magnetic field lines always point from the north pole of a magnet to the south pole.

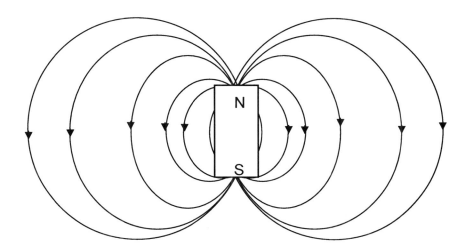

Magnetic field lines can be plotted with a magnetized needle that is free to turn in 3 dimensions. Usually a compass needle is used in demonstrations. The direction tangent to the magnetic field line is the direction the compass needle will point in a magnetic field. Iron filings spread on a flat surface or magnetic field viewing film which contains a slurry of iron filings are another way to see magnetic field lines.

When two magnets are placed close to one another magnetic field lines form between the north and south pole of each magnet individually and the field line also interact between the two magnets. The magnetic field in the area between the two magnets will depend on how the magnets are oriented with respect to one another. If the north pole of one magnet is placed near the south pole of the other magnet, for instance, you can see field lines starting at that north pole and ending at the south pole of the other magnet. If additional magnets and metal objects are placed in the arrangement the interactions become more complicated and various patterns are formed by the magnetic field lines.

Skill 20.3   Discuss factors that affect the strength of an electromagnet

The movement of electric charges (i.e., a current density **J**) results in a magnetic field **H**, as described by the Maxwell equation based on Ampere's law:

$$\nabla \times \mathbf{H}(r,t) = \frac{\partial \mathbf{D}(r,t)}{\partial t} + \mathbf{J}(r,t)$$

For the case where the time derivative of the electric flux density **D** is zero, this reduces to the simpler Ampere's law:

$$\nabla \times \mathbf{H}(r,t) = \mathbf{J}(r,t)$$

Electromagnets take advantage of this relationship between the current and the magnetic field by, for example, coiling a current-carrying wire (a solenoid), sometimes around a ferromagnetic or paramagnetic material. This creates a magnetic dipole with a strength that varies depending on a number of factors.

The direction of the dipole relative to the direction of current flow is determined by the right hand rule: the fingers curl in the direction of the current, and the thumb points in the (north) direction of the dipole.

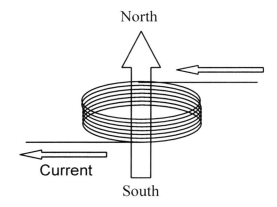

The strength of the electromagnet (i.e., the induced magnetic dipole) can be increased by adding to the number of turns (or loops) of current-carrying wire. The principle of superposition applies here and the induced magnetic field varies proportionally according to the number of turns. Also, by varying the current, the strength of the electromagnet can be increased or decreased. Imperfections in the solenoid, such as imperfectly packed wire loops, which can cause magnetic flux leakage, can adversely affect the strength of the electromagnet. Also, the spatial extent of the solenoid (or other form of electromagnet, such as a toroid) must be considered if exact numerical calculations are pursued.

Another critical factor that affects the electromagnet is the material used inside the coil; specifically, the magnetic properties of the material. Magnetizable materials with permeabilities (μ) greater than unity can be used to increase the magnitude of the induced magnetic dipole. Ferromagnetic materials, for example, can have very large permeabilities. One such example is iron. A ferromagnetic core is magnetized when the magnetic field induced by the current-carrying wire is applied resulting in a stronger magnetic dipole than would have been produced if a non-magnetic material (such as air) had been used. The ability of a magnetic material core to increase the strength of the electromagnet is limited, however. In the case of a ferromagnetic material, once all the magnetic domains have been aligned with the magnetic field of the electromagnet, saturation has been reached and the material cannot be magnetized further.

Skill 20.4   Magnitude and direction of the force on a charge or charges moving in a magnetic field

The magnetic force exerted on a charge moving in a magnetic field depends on the size and velocity of the charge as well as the magnitude of the magnetic field. One important fact to remember is that only the velocity of the charge in a direction perpendicular to the magnetic field will affect the force exerted. Therefore, a charge moving parallel to the magnetic field will have no force acting upon it whereas a charge will feel the greatest force when moving perpendicular to the magnetic field.

The direction of the magnetic force, or the magnetic component of the **Lorenz force** (force on a charged particle in an electrical and magnetic field), is always at a right angle to the plane formed by the velocity vector v and the magnetic field B and is given by applying the right hand rule - if the fingers of the right hand are curled in a way that seems to rotate the v vector into the B vector, the thumb points in the direction of the force. The magnitude of the force is equal to the cross product of the velocity of the charge with the magnetic field multiplied by the magnitude of the charge.

$$F = q \, (\mathbf{v} \times \mathbf{B}) \quad or \quad F = q \, v \, B \sin(\theta)$$

Where θ is the angle formed between the vectors of velocity of the charge and direction of magnetic field.

Problem: Assuming we have a particle of 1 x 10⁻⁶ kg that has a charge of -8 coulombs that is moving perpendicular to a magnetic field in a clockwise direction on a circular path with a radius of 2 m and a speed of 2000 m/s, let's determine the magnitude and direction of the magnetic field acting upon it.

Solution: We know the mass, charge, speed, and path radius of the charged particle. Combining the equation above with the equation for centripetal force we get

$$qvB = \frac{mv^2}{r} \quad \text{or} \quad B = \frac{mv}{qr}$$

Thus B= (1 x 10⁻⁶ kg) (2000m/s) / (-8 C)(2 m) = 1.25 x 10⁻⁴ Tesla

Since the particle is moving in a clockwise direction, we use the right hand rule and point our fingers clockwise along a circular path in the plane of the paper while pointing the thumb towards the center in the direction of the centripetal force. This requires the fingers to curl in a way that indicates that the magnetic field is pointing out of the page. However, since the particle has a negative charge we must reverse the final direction of the magnetic field into the page.

A **mass spectrometer** measures the mass to charge ratio of ions using a setup similar to the one described above. m/q is determined by measuring the path radius of particles of known velocity moving in a known magnetic field.

A **cyclotron**, a type of particle accelerator, also uses a perpendicular magnetic field to keep particles on a circular path. After each half circle, the particles are accelerated by an electric field and the path radius is increased. Thus the beam of particles moves faster and faster in a growing spiral within the confines of the cyclotron until they exit at a high speed near the outer edge. Its compactness is one of the advantages a cyclotron has over linear accelerators.

The **force on a current-carrying conductor** in a magnetic field is the sum of the forces on the moving charged particles that create the current. For a current I flowing in a straight wire segment of length l in a magnetic field B, this force is given by

$$F = Il \times B$$

where l is a vector of magnitude l and direction in the direction of the current.

When a current-carrying loop is placed in a magnetic field, the net force on it is zero since the forces on the different parts of the loop act in different directions and cancel each other out. There is, however, a net torque on the loop that tends to rotate it so that the area of the loop is perpendicular to the magnetic field. For a current I flowing in a loop of area A, this torque is given by

$$\tau = IA\hat{n} \times \boldsymbol{B}$$

where $\hat{n}$ is the unit vector perpendicular to the plane of the loop.

Skill 20.5   Discuss the use of electromagnetism in technology (e.g., motors, generators, meters)

Electromagnetism is the foundation for a vast number of modern technologies ranging from computers to communications equipment. More mundane technologies such as motors and generators are also based upon the principles of electromagnetism. The particular understanding of electrodynamics can either be in terms of quantum mechanics (quantum electrodynamics) or classical electrodynamics, depending on the type of phenomenon being analyzed. In classical electrodynamics, which is a sufficient approximation for most situations, the electric field, resulting from electric charge, and the magnetic field, resulting from moving charges, are the parameters of interest and are related through Maxwell's equations.

Motors
Electric motors are found in many common appliances such as fans and washing machines. The operation of a motor is based on the principle that a magnetic field exerts a force on a current carrying conductor. This force is essentially due to the fact that the current carrying conductor itself generates a magnetic field; the basic principle that governs the behavior of an electromagnet. In a motor, this idea is used to convert **electrical energy into mechanical energy**, most commonly rotational energy. Thus the components of the simplest motors must include a strong magnet and a current-carrying coil placed in the magnetic field in such a way that the force on it causes it to rotate.

A typical motor is composed of a stationary portion, called the stator, and a rotating (or moving) portion, called the rotor. Coils of wire that serve as electromagnets are wound on the armature, which can be either the stator or the rotor, and are powered by an electric source. Motors use electric current to generate a magnetic field around an electromagnet, which results in a rotational force due to the presence of an external magnetic field (either from permanent magnets or electromagnets). The designs of various motors can differ dramatically, but the general principles of electromagnetism that describe their operation are generally the same.

Generators

Generators are in effect "reverse motors". They exploit electromagnetic induction to generate electricity. Thus, if a coil of wire is rotated in a magnetic field, an alternating EMF is produced which allows current to flow. Any number of energy sources can be used to rotate the coil, including combustion, nuclear fission, flowing water or other sources. Therefore, generators are devices that convert **mechanical or other forms of energy into electrical energy**. For more about generators see **Skill 21.4**.

Meters

A number of different types of meters use electromagnetism or are designed to measure certain electromagnetic parameters. For example, older forms of ammeters (galvanometers), when supplied with a current, provided a measurement through the deflection of a spring-loaded needle. A coil connected to the needle acted as an electromagnet which, in the presence of a permanent magnetic field, would be deflected in the same manner as a rotor, as mentioned previously. The strength of the electromagnet, and thus the extent of the deflection, is proportional to the current. Further, the spring limits the deflection in such a manner that a reasonably accurate measurement of the current is provided.

Magnetic Media

Although the cassette tape has fallen out of favor with popular culture, magnetic media are still widely used for information storage. Magnetic strips on the back of credit and identification cards, computer hard drive disks and magnetic tapes (such as those contained in cassettes) are all examples of magnetic media. The principle underlying magnetic media is the sequential magnetization of a region of the medium. A special head is able to detect spatial magnetic fluctuations in the medium which are then converted into an electrical signal. The head is often able to "write" to the medium as well. The electrical signal from the medium can be converted into sound or video, as with the video or audio cassette player, or it can be digitized for use in a computer.

COMPETENCY 21.0 UNDERSTAND AND APPLY THE PRINCIPLES OF ELECTROMAGNETIC INDUCTION AND AC CIRCUITS

Skill 21.1 Analyzing factors that affect the magnitude of an induced electromotive force (EMF)

When the magnetic flux through a coil is changed, a voltage is produced which is known as induced electromagnetic force. Magnetic flux is a term used to describe the number of magnetic fields lines that pass through an area and is described by the equation:

$$\Phi = B\,A\,\cos\theta$$

Where Φ is the angle between the magnetic field B, and the normal to the plane of the coil of area A

By changing any of these three inputs, magnetic field, area of coil, or angle between field and coil, the flux will change and an EMF can be induced. The speed at which these changes occur also affects the magnitude of the EMF, as a more rapid transition generates more EMF than a gradual one. This is described by **Faraday's law** of induction:

$$\varepsilon = -N\,\Delta\Phi\,/\,\Delta t$$

where ε is emf induced, N is the number of loops in a coil, t is time, and Φ is magnetic flux

The negative sign signifies **Lenz's law** which states that induced emf in a coil acts to oppose any change in magnetic flux. Thus the current flows in a way that creates a magnetic field in the direction opposing the change in flux. See the next section for the right-hand rule that determines the direction of the induced current.

Example:
Consider a coil lying flat on the page with a square cross section that is 10 cm by 5 cm. The coil consists of 10 loops and has a magnetic field of 0.5 T passing through it coming out of the page. Let's find the induced EMF when the magnetic field is changed to 0.8 T in 2 seconds.

First, let's find the initial magnetic flux: Φ_i
$\Phi_i = BA \cos\theta = (.5\ \text{T})\,(.05\ \text{m})\,(.1\text{m}) \cos 0° = 0.0025\ \text{T m}^2$
And the final magnetic flux: Φ_f
$\Phi_f = BA \cos\theta = (0.8\ \text{T})\,(.05\ \text{m})\,(.1\text{m}) \cos 0° = 0.004\ \text{T m}^2$
The induced emf is calculated then by
$\varepsilon = -N\,\Delta\Phi\,/\,\Delta t = -\,10\,(.004\ \text{T m}^2 - .0025\ \text{T m}^2)\,/\,2\ \text{s} = -0.0075\ \text{volts.}$

Skill 21.2 Applying the appropriate hand rule to determine the direction of an induced current

The negative sign in Faraday's Law (described in the previous section) leads to Lenz's law which states that the induced current must produce a magnetic field that opposes the change in the applied magnetic field. This is simply an expression of the conservation of energy.

The right- or left-hand rule applies only in the case of a particular current convention: the right-hand rule is typically applied in the case of the positive current convention (i.e., the direction of flow of positive charge). Since positive and negative current are complementary, however, the rules for negative current simply involve using the left hand instead of the right hand. For simplicity, only the right-hand rule will be discussed explicitly.

The right-hand rule states that if the fingers of the right hand are curled in the direction of the positive current flow, the resulting magnetic flux is in the direction of the extended thumb. According to Lenz's law, when the magnetic flux through the surface bounded by the conductor increases, the induced current, $i_{induced}$, must produce a magnetic flux that is in the opposite direction to that of the applied flux. This is illustrated below.

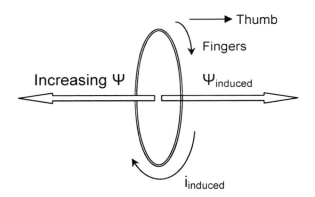

When the applied magnetic flux through the surface decreases, the direction of the induced flux must likewise coincide with the direction of the applied flux.

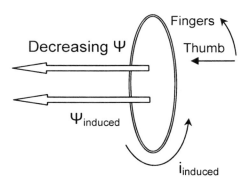

Thus, the induced magnetic flux must oppose the change in the applied flux. The direction of the induced positive current is the same as the direction of the curled fingers of the right hand when the thumb is extended in the direction of the induced magnetic flux. If negative current is of interest (for example, the flow of electrons), then the left-hand must be substituted in place of the right hand; the thumb follows the same rules, but the direction of the curled fingers indicates the direction of negative current flow.

Skill 21.3   Analyzing an AC circuit, including relationships involving impedance, reactance, and resonance

Alternating current (AC) is a type of electrical current with cyclically varying magnitude and direction. This is differentiated from direct current (DC), which has constant direction. AC is the type of current delivered to businesses and residences.

Though other waveforms are sometimes used, the vast majority of AC current is sinusoidal. Thus we can use wave terminology to help us describe AC current. Since AC current is a function of time, we can express it mathematically as:

$$v(t) = V_{peak} \cdot \sin(\omega t)$$

where V_{peak}= the peak voltage; the maximum value of the voltage
ω=angular frequency; a measure of rotation rate
t=time

A few more terms are useful to help us characterize AC current:

Peak-to-peak value: The difference between the positive and negative peak values. Thus peak-to-peak value is equal to 2 x V_{peak}.

Root mean square value (V_{rms}, I_{rms}): A specific type of average found by the following formulae:

$$V_{rms} = \frac{V_{peak}}{\sqrt{2}} \quad ; \quad I_{rms} = \frac{I_{peak}}{\sqrt{2}} ; \quad I_{rms} = \frac{V_{rms}}{R}$$

V_{rms} is useful because an AC current will deliver the same power as a DC current if its $V_{rms} = V_{DC}$, i.e. average power

$$P_{av} = V_{rms} I_{rms}.$$

Frequency: Describes how often the wave passes through a particular point per unit time. Note that this is physical frequency, f, which is related to the angular frequency ω by:

$$\omega = 2\pi f$$

Impedance: A measure of opposition to an alternating current. It is similar to resistance and also has the unit ohm. However, due to the phased nature of AC, impedance is a complex number, having both real and imaginary components. The resistance is the real part of impedance while the reactance of capacitors and inductors constitute the imaginary part.

Resonant frequency: The frequency at which the impedance between the input and output of the circuit is minimum. At this frequency a phenomenon known as electrical resonance occurs.

Reactance: The impedance contributed by inductors and capacitors in AC circuit. Mathematically, reactance is the imaginary part of impedance. The relationship between impedance (Z), resistance(R), and reactance (X) is given by below.

$$Z = R + Xi$$

1. Remember that i= $\sqrt{-1}$

Problem:
An AC current has V_{rms}=220 V. What is its peak-to-peak value?

Solution:
We simply determine V_{peak} from the definition of V_{rms}:

$$V_{rms} = \frac{V_{peak}}{\sqrt{2}}$$

$$V_{peak} = V_{rms} \times \sqrt{2} = 220V \times \sqrt{2} = 311.12V$$

Therefore,

$$V_{peak-to-peak} = 2 \times V_{peak} = 2 \times 311.12V = 622.24V$$

Skill 21.4   Analyzing the functions of transformers and generators

Transformers and generators have critical functions in the production and use of electrical power. A generator is essentially a transducer that converts one form of energy, such as mechanical or heat energy, into electrical energy. It uses electromagnetic induction to generate electricity. The simplest way to implement a generator is to rotate a wire loop in a permanent magnetic field.

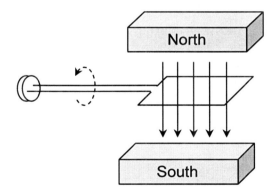

Depending on the strength of the magnetic field and the rate of rotation of the loop, a voltage of a specific amplitude can be produced. Additionally, the rotation of the wire loop means that the magnetic flux is at times increasing and at times decreasing, resulting in a sinusoidal voltage with a frequency equal to the rotation frequency of the loop. Electricity supplied by public utilities, for example, is 120 V at a frequency of 60 Hz. The source of energy for the wire loop rotator (turbine) can be a dammed river or an engine run by combustion, a nuclear reaction or another process. The energy extracted from these sources (either as mechanical or heat energy) is transferred to electrical energy through the use of the generator.

Transformers fulfill another purpose in the context of electrical power by either stepping up or stepping down the voltage (and, concomitantly, stepping down or stepping up the current). They do this by magnetically coupling two circuits together. This allows the transfer of energy between these two circuits without requiring motion. Typically, a transformer consists of a couple of coils and a magnetic core. A changing voltage applied to one coil (the primary inductor) creates a flux in the magnetic core, which induces voltage in the other coil (the secondary inductor).

One of the simplest ways to control the flux linkage between the coils is by appropriately increasing or decreasing the number of turns in the inductors relative to one another. In the ideal case, where all the flux produced by the first inductor (L_1 above) is linked to the second inductor (L_2 above), the voltage across L_2 (V_t) is equal to the voltage across L_1 (V) multiplied by the ratio of the number of turns in L_2 (N_2) to the number of turns in L_1 (N_1).

$$V_t = V \frac{N_2}{N_1}$$

Thus, if there are twice as many turns in L_2 as in L_1, twice the flux is linked to L_2 and the voltage is doubled. In reality, however, not all the flux is linked to the second inductor. The use of ferromagnetic cores, especially if a loop of ferromagnetic material is used such that it serves as the core for both inductors, increases the flux linkage and brings the transformer closer to the ideal case. All transformers operate on this simple principle though they range in size and function from those in tiny microphones to those that connect the components of the US power grid.

As a side note, any circuit (which is a wire loop of one form or another) has an inherent inductance that can link the flux from another circuit. Thus, "crosstalk" between circuits can be an undesirable effect that results from the same principles that are used to intentionally design the beneficial transformer device.

Problem: If a step-up transformer has 500 turns on its primary coil and 800 turns on its secondary coil, what will be the output (secondary) voltage be if the primary coil is supplied with 120 V?

Solution:

$$\frac{V_s}{V_p} = \frac{n_s}{n_p}$$

$$V_s = \frac{n_s}{n_p} \times V_p = \frac{800}{500} \times 120V = 192V$$

COMPETENCY 22.0   UNDERSTAND THE PRINCIPLES OF CONDUCTORS, SEMICONDUCTORS, AND SUPERCONDUCTORS

Skill 22.1   Conductors, semiconductors, and superconductors

Semiconductors, conductors and superconductors are differentiated by how "easily" current can flow in the presence of an applied electric field. Dielectrics (insulators) conduct little or no current in the presence of an applied electric field, as the charged components of the material (for example, atoms and their associated electrons) are tightly bound and are not free to move within the material. In the case of materials with some amount of conductivity, so-called valence (higher energy level) electrons are only loosely bound and may move among positive charge centers (i.e., atoms or ions).

Conductors, such as metals like aluminum and iron, have numerous valence electrons distributed among the atoms that compose the material. These electrons can move freely, especially under the influence of an applied electric field. In the absence of any applied field, these electrons move randomly, resulting in no net current. If a static electric field is applied, the mobile electrons reorient themselves to minimize the energy of the system and create an equipotential on the surface of the metal (in the ideal case of infinite conductivity).The highly mobile charge carriers do not allow a net field inside the metal; thus conductors can "shield" electromagnetic fields.

Semiconductors, such as silicon and germanium, are materials that can neither be described as conductors, nor as insulators. **A certain number of charge carriers are mobile in the semiconductor**, but this number is nowhere near the free charge populations of conductors such as metals. "Doping" of a semiconductor by adding so-called donor atoms or acceptor atoms to the intrinsic (or pure) semiconductor can increase the conductivity of the material. Furthermore, combination of a number of differently doped semiconductors (such as donor-doped silicon and acceptor-doped silicon) can produce a device with beneficial electrical characteristics (such as the diode). The conductivity of such devices can be controlled by applying voltages across specific portions of the device.

A useful way to visualize the relationship of conductors and semiconductors is by way of energy band diagrams. For purposes of comparison, an example of an insulator is included here as well.

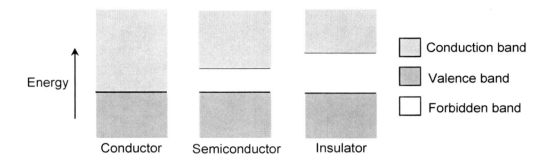

According to the theory of energy bands, as derived from quantum mechanics, electrons in the ground state reside in the valence band and are bound to their associated atoms or molecules. If the electrons gain sufficient energy, such as through heat, they can jump across the forbidden band (in which no electrons may exist) to the conduction band, thus becoming free electrons that can form a current in the presence of an applied field. For conductors, the valence band and conduction band meet or overlap, allowing electrons to easily jump to unoccupied conduction states. Thus, conductors have an abundance of free electrons. It is noteworthy that only two bands are shown in the diagram above, but that an infinite number of bands may exist at higher energies. A more general statement of the difference in band structures is that, for insulators and semiconductors, the valence electrons fill up all the states in a particular band, leaving a gap between the highest energy valence electrons and the next available band. The difference between these two types of materials is simply a matter of the "size" of the forbidden band. For conductors, the band that contains the highest energy electrons has additional available states.

A nearly ideal conducting material is a superconductor. As a material increases in temperature, increased vibrational motion of the atoms or molecules leads to decreased charge carrier mobility and decreased conductivity. In the case of semiconductors, the increase in free carrier population outweighs the loss in mobility of the charge carriers, meaning that the semiconductor increases in conductivity as temperature increases. Superconductors, on the other hand, reach their peak conductivity at extremely low temperatures (although there are currently numerous efforts to achieve superconductivity at higher and higher temperatures, with room temperature or higher being the ultimate goal).

The critical temperature of the material is the temperature at which superconducting properties emerge. At this temperature, the material has a nearly infinite conductivity and maintains an almost perfect equipotential across its surface when in the presence of a static electric field. Inside a superconductor, the electric field is virtually zero at all times. As a result, the time derivative of the electric flux density is zero, and, by Maxwell's equations, the magnetic flux density must likewise be zero.

$$\nabla \times \mathbf{H} = \frac{\partial \mathbf{D}}{\partial t} + \mathbf{J}$$

Since the electric field is also zero, the current density $\mathbf{J}$ inside the superconductor must also be zero (or very nearly so). **This elimination of the magnetic flux density inside a superconductor is called the Meissner effect**.

Skill 22.2 Analyze current-voltage characteristics of typical solid state diodes and Zener diodes and explain the function of a diode in a given circuit

A diode is an electrical aspect of a circuit that serves to maintain the continual flow of electricity in a single direction, while blocking electron flow in the opposite direction. Today, the most common diodes are made from semiconductor materials such as silicon or germanium. Most modern diodes are based on semiconductor p-n junctions. In a p-n diode, the forward-headed conventional current can flow from the p-type side (the anode) to the n-type side (the cathode), but cannot flow in the opposite direction.

A conventional solid-state diode will not allow a significant amount of current to flow in the reverse direction if the current is below its reverse breakdown voltage. When the reverse bias breakdown voltage is exceeded, however, a conventional diode is exposed to high current flow due to **avalanche breakdown**. Avalanche breakdown occurs within insulating or semiconducting structures when the electric field in the material is great enough to accelerate free electrons to the point that, when they strike fixed atoms in the material, they can knock other electrons free so that the number of free electrons is thus increased rapidly as newly generated particles became part of the process. Unless the current is blocked by external circuitry, the diode will be permanently damaged. In situations involving a large forward current flow in the direction of the circuit, the diode will exhibit a voltage drop due to built-in voltage and internal resistance. The amount of the voltage drop depends on the semiconductor material and the doping concentrations.

Zener diodes operate somewhat differently except that they can allow the passage of electrons in the opposite direction of the current when the Zener voltage is exceeded. These devices have a greatly-reduced breakdown voltage and contain a heavily doped p-n junction so that electrons can tunnel from the valence band of the p-material to the conduction band of the n-material. When the electron flow is reversed, there is a controlled breakdown so that the voltage across the diode is equal to the Zener voltage and the diode is not damaged.

Electricity wastes a little energy pushing its way through the diode, similar to the way an individual pushes open a door with a spring. This means that there is a small voltage across a conducting diode, it is called the **forward voltage drop** and is about 0.7V for all normal diodes which are made from silicon. The forward voltage drop of a diode is almost constant whatever the current passing through the diode giving them a very steep relationship (see the current-voltage graph).

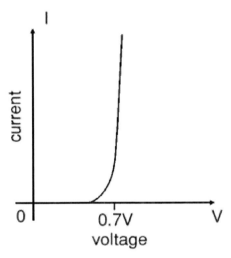

Voltage and Current Relationship in a Silicon Diode

Skill 22.3   Compare NPN and PNP transistors and identify correct terminal connections in a given circuit

The transistor is considered by many to be the greatest invention of the twentieth century. It is found as the key component in practically all modern electronic devices, including televisions and computers. Transistors have the function of being amplifiers of current, as well as being used in digital and analog functions, including switching, voltage regulation, signal modulation, and oscillation. Transistors are in/out devices in a circuit that amplify the current. Transistors consist of three terminals: the source, the gate, and the drain. Transistors may be packaged individually or as part of an integrated circuit chip.

There are two main types of transistors: Bipolar Junction Transistors (BJTs) and Field Effect Transistors (FETs). Current is controlled through the application of current into the transistors, which amplifies or increases the current between the input and output of the transistors. The amount of current flowing across the transistor depends on the type of the transistor. The two types of Bipolar Junction Transistors are the NPN transistor and the PNP transistor.

The NPN transistor is the most common type of transistor in use today. Silicon containing boron impurities is called P-type silicon for "positive" since it lacks electrons (cathode). Silicon containing phosphorus impurities is called N-type silicon for "negative" since it has excess free electrons (anode). In the NPN-type transistor, both the source and the drain are negatively charged and sit on a positively charged well of P-silicon. The NPN transistor consists of a layer of P-doped semiconductor, called the base that sits below the two N-doped layers. A small current enters the base and is amplified in the collector output. When positive voltage is applied to the gate, electrons in the p-silicon are attracted to the area under the gate, forming an electron channel between the source and the drain. When positive voltage is applied to the drain, the electrons are pulled from the source to the drain. In this state the transistor is on.

PNP transistors provide less performance than NPN transistors and are less commonly used. The device consists of a N-doped semiconductor which act as a base with two layers of P-doped material above it. The collector is operated at ground and the emitter is connected to a positive voltage via an electric load. A small current flowing from the base allows for a much greater current to flow from the emitter to the collector.

In **transistor circuits**, the circuit is labeled in the following way:

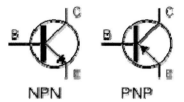

NPN   PNP

B stands for the base, E is the emitter and C is the collector.

The current enters the base of the transmitter in a NPN transmitter and exits via the emitter. The collector uses augmentation of the voltage entering the base to brightly light an LED light or to provide high voltage to a particular aspect of the circuit.

Skill 22.4   Identify the function of a transistor in a given electrical circuit

As explained above, transistors can serve a variety of purposes. The use of the transistor depends partially upon the state in which it operates. Transistors in "cut off" have no collector current and are useful in switch operation. In the active region, there is some collector current (more than a few tenths of a volt above the emitter) and the transistor is more than likely being used as an amplifier. In saturation, the collector is a few tenths of a volt above the emitter and the transistor may be used in "switch on". Simple circuits are diagramed below. In each one, the transistor serves a difference purpose which is described.

Transistor as a switch
In this circuit, the transistor is functioning as a switch, controlling power delivery to R_1. A mechanical switch is also present in this circuit, as this is a common arrangement. Closing this mechanical switch drives sufficient base current to force the voltage of the collector below that of the base. Note that the resistance value R_2 must be within an appropriate range for this to happen. The collector voltage will be close to that of the

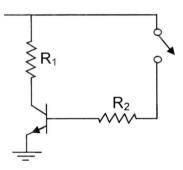

emitter (ground) and the transistor will be in saturation. At this point, the transistor switch is "on"; increasing the base current cannot trigger a further increase in the collector current. If the mechanical switch is opened again, the voltage drop from base to emitter will again decrease (cut-off) and the transistor will shut off power to R_1.

Transistor as a logic gate

This circuit is similar in form to the one above; however, here the input (V_{in}) and output (V_{out}) voltage drops are different. This circuit is a rudimentary "NOT" logic gate which inverts the input. That is, when V_{in} is high, V_{out} will be low and when V_{in} is low, V_{out} will be high. This happens when V_{in} is above a threshold, the base current turns on and current is directed through R_1. When V_{in} drops below the threshold, the base and collector currents are zero and the current cannot flow through R_1 and is redirected, thus increasing V_{out}.

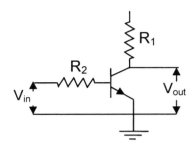

Transistor as an amplifier

This circuit is a common emitter amplifier, in which the input voltage (V_{in}) is multiplied by a constant voltage gain (A_v) to produce an output voltage (V_{out}). Note that this means that V_{out} is linearly proportional to V_{in}. In amplifiers, resistor values are chosen such that a small current always flows from the collector to the emitter, keeping it in the quiescent state (the active region). Thus any change in V_{in} is mirrored by the emitter. This behavior is sometimes explained as being analogous to a valve. The small current in the base serves as the "valve" and controls the larger current that flows from collector to emitter. Thus the "signal" (variation in the base current) is reproduced as an amplified variation in the current flowing from the collector to the emitter.

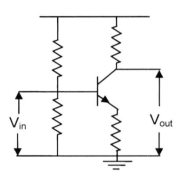

COMPETENCY 23.0   UNDERSTAND WAVES AND WAVE MOTION AND SOLVE PROBLEMS INVOLVING WAVE MOTION

Skill 23.1   Understand waves and wave motion

To fully understand waves, it is important to understand many of the terms used to characterize them.

Wave velocity: Two velocities are used to describe waves. The first is phase velocity, which is the rate at which a wave propagates. For instance, if you followed a single crest of a wave, it would appear to move at the phase velocity. The second type of velocity is known as group velocity and is the speed at which variations in the wave's amplitude shape propagate through space. Group velocity is often conceptualized as the velocity at which energy is transmitted by a wave. Phase velocity is denoted v_p and group velocity is denoted v_g. In a medium with refractive index independent of frequency, such as vacuum, the phase velocity is equal to the group velocity.

Crest: The maximum value that a wave assumes; the highest point.

Trough: The lowest value that a wave assumes; the lowest point.

Nodes: The points on a wave with minimal amplitude.

Antinodes: The farthest point from the node on the amplitude axis; both the crests and the troughs are antinodes.

Amplitude: The distance from the wave's highest point (the crest) to the equilibrium point. This is a measure of the maximum disturbance caused by the wave and is typically denoted by A.

Wavelength: The distance between any two sequential troughs or crests denoted λ and representing a complete cycle in the repeated wave pattern.

Period: The time required for a complete wavelength or cycle to pass a given point. The period of a wave is usually denoted T.

Frequency: The number of periods or cycles per unit time (usually a second). The frequency is denoted f and is the inverse of the wave's period (that is, $f=1/T$).

Phase: This is a given position in the cycle of the wave. It is most commonly used in discussing a "being out of phase" or a "phase shift", an offset between waves.

We can visualize several of these terms on the following diagram of a simple, periodic sine wave on a scale of distance displacement (x-axis) vs. (y-axis):

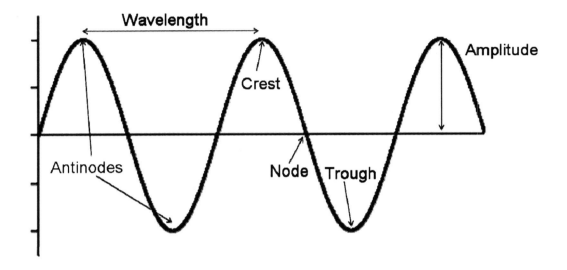

This diagram of a sinusoidal wave shows us displacement caused by the wave as it propagates through a medium. This displacement can be graphed against either time or distance. Note how displacement depends on the distance which the wave has traveled/ how much time has elapsed. So if we chose a particular displacement (let's say the crest), the wave will return to that displacement value (i.e., crest again) after one period (T) or one wavelength (λ).

Skill 23.2 Applying the wave equation to determine a wave's velocity, wavelength, or frequency

The general equation for a wave is a partial differential equation which can be simplified to express the behavior of commonly encountered harmonic waveforms (See **Skill 16.4**). For instance, for a standing wave:

$$y(z,t) = A(z,t)\sin(kz - \omega t + \phi)$$

where y=displacement
z=distance
t=time
k=wave number = $2\pi/\lambda$
ω=angular frequency= $2\pi f$
ϕ=phase
A(z,t)=the amplitude envelope of the wave

The important concept to note is that y is a function of both z and t. This means that the wave's position depends on both time and distance, just as was seen in the diagram above.

The phase velocity of a wave is related to its wavelength and frequency. Taking light waves, for instance, the speed of light c is equal to the distance traveled divided by time taken. Since the light wave travels the distance of one wavelength λ in the period of the wave T,

$$c = \frac{\lambda}{T}$$

The frequency of a wave, f, is the number of completed periods in one second. In general,

$$f = \frac{1}{T}$$

So the formula for the speed of light can be rewritten as

$$c = \lambda f$$

Thus the phase velocity of a wave is equal to the wavelength times the frequency.

COMPETENCY 24.0   APPLY THE PRINCIPLES OF WAVE REFLECTION, REFRACTION, DIFFRACTION, INTERFERENCE, POLARIZATION, DISPERSION, AND THE DOPPLER EFFECT

Skill 24.1   Demonstrate applications of wave reflection, refraction, diffraction, interference, polarization, and dispersion

Wave **refraction** is a change in direction of a wave due to a change in its speed. This most commonly occurs when a wave passes from one material to another, such as a light ray passing from air into water or glass. However, light is only one example of refraction; any type of wave can undergo refraction. Another example would be physical waves passing from water into oil. At the boundary of the two media, the wave velocity is altered, the direction changes, and the wavelength increases or decreases. However, the frequency remains constant.

The **index of refraction**, n, is the amount by which light slows in a given material and is defined by the formula

$$n = \frac{c}{v}$$

where v represents the speed of light through the given material.

Problem:
The speed of light in an unknown medium is measured to be $1.24x10^8 m/s$. What is the index of refraction of the medium?

Solution:

$$n = \frac{c}{v}$$

$$n = \frac{3.00x10^8}{1.24x10^8} = 2.42$$

Referring to a standard table showing indices of refraction, we would see that this index corresponds to the index of refraction for diamond.

Reflection is the change in direction of a wave at an interface between two dissimilar media such that the wave returns into the medium from which it originated. The most common example of this is light waves reflecting from a mirror, but sound and water waves can also be reflected. **The law of reflection states that the angle of incidence is equal to the angle of reflection.**

Diffraction occurs when part of a wave front is obstructed. Diffraction and interference are essentially the same physical process. Diffraction refers to various phenomena associated with wave propagation such as the bending, spreading, and interference of waves emerging from an aperture. It occurs with any type of wave including sound waves, water waves, and electromagnetic waves such as light and radio waves.

Here, we take a close look at important phenomena like single-slit diffraction, double-slit diffraction, diffraction grating, other forms of diffraction and lastly interference.

1. Single-slit diffraction: The simplest example of diffraction is single-slit diffraction in which the slit is narrow and a pattern of semi-circular ripples is formed after the wave passes through the slit.

2. Double-slit diffraction: These patterns are formed by the interference of light diffracting through two narrow slits.

3. Diffraction grating: Diffraction grating is a reflecting or transparent element whose optical properties are periodically modulated. In simple terms, diffraction gratings are fine parallel and equally spaced grooves or rulings on a material surface. When light is incident on a diffraction grating, light is reflected or transmitted in discrete directions, called diffraction orders. Because of their light dispersive properties, gratings are commonly used in monochromators and spectrophotometers. Gratings are usually designated by their groove density, expressed in grooves/millimeter. A fundamental property of gratings is that the angle of deviation of all but one of the diffracted beams depends on the wavelength of the incident light.

4. Other forms of diffraction:

i) Particle diffraction: It is the diffraction of particles such as electrons, which is used as a powerful argument for quantum theory. It is possible to observe the diffraction of particles such as neutrons or electrons and hence we are able to infer the existence of wave particle duality.

ii) Bragg diffraction: This is diffraction from a multiple slits, and is similar to what occurs when waves are scattered from a periodic structure such as atoms in a crystal or rulings on a diffraction grating. Bragg diffraction is used in X-ray crystallography to deduce the structure of a crystal from the angles at which the X-rays are diffracted from it.

5. Interference: Interference is described as the superposition of two or more waves resulting in a new wave pattern. Interference is involved in Thomas Young's double slit experiment where two beams of light which are coherent with each other interfere to produce an interference pattern. Light from any source can be used to obtain interference patterns. For example, Newton's rings can be produced with sun light. However, in general, white light is less suited for producing clear interference patterns as it is a mix of a full spectrum of colors. Sodium light is close to monochromatic and is thus more suitable for producing interference patterns. The most suitable is laser light as it is almost perfectly monochromatic.

Dispersion is the separation of a wave into its constituent wavelengths due to interaction with a material occurring in a wavelength-dependent manner (as in thin-film interference for instance). Dispersive prisms separate white light into these constituent colors by relying on the differences in refractive index that result from the varying frequencies of the light. Prisms rely on the fact that light changes speed as it moves from one medium to another. This then causes the light to be bent and/or reflected. The degree to which bending/reflection occurs is a function of the light's angle of incidence and the refractive indices of the media.

Polarization is a property of transverse waves that describes the plane perpendicular to the direction of travel in which the oscillation occurs. In unpolarized light, the transverse oscillation occurs in all planes perpendicular to the direction of travel. Polarized light (created, for instance, by using polarizing filters that absorb light oscillating in other planes) oscillates in only a selected plane. An everyday example of polarization is found in polarized sunglasses which reduce glare.

Skill 24.2   Explain and give examples of use of the Doppler effect

The **Doppler effect** is the name given to the perceived change in frequency that occurs when the observer or source of a wave is moving. Specifically, the perceived frequency increases when a wave source and observer move toward each other and decreases when a source and observer move away from each other. Thus, the source and/or observer velocity must be factored in to the calculation of the perceived frequency. The mathematical statement of this effect is:

$$f' = f_0 \left(\frac{v \pm v_o}{v \pm v_s} \right)$$

where f'= observed frequency
f_0= emitted frequency
v= the speed of the waves in the medium
v_s= the velocity of the source (positive in the direction away from the observer)
v_o= the velocity of the observer (positive in the direction towards the source)

Note that any motion that changes the perceived frequency of a wave will cause the Doppler effect to occur. Thus, the wave source, the observer position, or the medium through which the wave travels could possess a velocity that would alter the observed frequency of a wave.

You may view animations of stationary and moving wave sources at the following URL:

http://www.kettering.edu/~drussell/Demos/doppler/doppler.html

So, let's consider two examples involving sirens and analyze what happens when either the source or the observer moves. First, imagine a person standing on the side of a road and a police car driving by with its siren blaring. As the car approaches, the velocity of the car will mean sound waves will "hit" the observer as the car comes closer and so the pitch of the sound will be high. As it passes, the pitch will slide down and continue to lower as the car moves away from the observer. This is because the sound waves will "spread out" as the source recedes. Now consider a stationary siren on the top of fire station and a person driving by that station. The same Doppler effect will be observed: the person would hear a high frequency sound as he approached the siren and this frequency would lower as he passed and continued to drive away from the fire station.

The Doppler effect is observed with all types of electromagnetic radiation. In everyday life, we may be most familiar with the Doppler effect and sound, as in the example above. However, we can observe example of it throughout the EM spectrum. For instance, the Doppler effect for light has been exploited by astronomers to measure the speed at which stars and galaxies are approaching. Another familiar application is the use of Doppler radar by police to detect the speed of oncoming cars.

Skill 24.3   Apply Snell's law to determine index of refraction, angle of incidence, angle of refraction, or critical angle

Snell's Law describes how light bends, or refracts, when traveling from one medium to the next. It is expressed as

$$n_1 \sin\theta_1 = n_2 \sin\theta_2$$

where n_i represents the index of refraction in medium i, and θ_i represents the angle the light makes with the normal in medium i.

<u>Problem:</u>   The index of refraction for light traveling from air into an optical fiber is 1.44. (a) In which direction does the light bend? (b) What is the angle of refraction inside the fiber, if the angle of incidence on the end of the fiber is 22°?

Solution: (a) The light will bend toward the normal since it is traveling from a rarer region (lower n) to a denser region (higher n).

(b) Let air be medium 1 and the optical fiber be medium 2:

$$n_1 \sin\theta_1 = n_2 \sin\theta_2$$
$$(1.00)\sin 22° = (1.44)\sin\theta_2$$
$$\sin\theta_2 = \frac{1.00}{1.44}\sin 22° = (.6944)(.3746) = 0.260$$
$$\theta_2 = \sin^{-1}(0.260) = 15°$$

The angle of refraction inside the fiber is $15°$.

Reflection may occur whenever a wave travels from a medium of a given refractive index to another medium with a different index. A certain fraction of the light is reflected from the interface and the remainder is refracted. However, when the wave is moving from a dense medium into one less dense, that is the refractive index of the first is greater than the second, a **critical angle** exists which will create a phenomenon known as **total internal reflection.** In this situation all of the wave is reflected. The critical angle of incidence θ_c is the one for which the angle of reflection is 90 degrees. Thus, according to Snell's law

$$n_1 \sin\theta_c = n_2 \sin 90^0 \Rightarrow \theta_c = \sin^{-1}\frac{n_2}{n_1}$$

Snell's law may also be used to understand the phenomenon of dispersion since it relates the angle of refraction at the boundary between two media to the relative refractive indices. Since different frequency components of visible light have different indices of refraction in any medium other than vacuum, each component of a beam of white light is refracted at a different angle when it crosses a surface resulting in a separation of the colors.

Skill 24.4   Solve problems involving diffraction and interference in single and multiple slits

All wavelengths in the EM spectrum can experience interference but it is easy to comprehend instances of interference in the spectrum of visible light. One classic example of this is **Thomas Young's double-slit experiment**. In this experiment a beam of light is shone through a paper with two slits and a striated wave pattern results on the screen. The light and dark bands correspond to the areas in which the light from the two slits has constructively (bright band) and destructively (dark band) interfered. Light from any source can be used to obtain interference patterns. For example, Newton's rings can be produced with sun light. However, in general, white light is less suited for producing clear interference patterns as it is a mix of a full spectrum of colors. Sodium light is close to monochromatic and is thus more suitable for producing interference patterns. The most suitable is laser light as it is almost perfectly monochromatic.

Problem: The interference maxima (location of bright spots created by constructive interference) for double-slit interference are given by

$$\frac{n\lambda}{d} = \frac{x}{D} = \sin\theta \quad n=1,2,3\ldots$$

where λ is the wavelength of the light, d is the distance between the two slits, D is the distance between the slits and the screen on which the pattern is observed and x is the location of the nth maximum. If the two slits are 0.1mm apart, the screen is 5m away from the slits, and the first maximum beyond the center one is 2.0 cm from the center of the screen, what is the wavelength of the light?

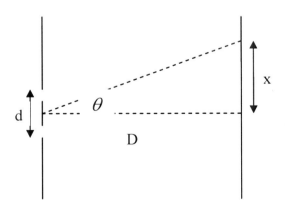

Solution: λ = xd/(Dn) = 0.02 x 0.0001/ (5 x1) = 400 nanometers

For **single-slit diffraction**, a central bright fringe of light is observed since waves from all points in the slit travel approximately the same distance to the center of the pattern and are in phase there. An interference pattern of alternate bright and dark fringes is observed around the central spot.

The location of the dark fringes is given by

$$a \sin\theta = m\lambda \quad \text{for } m = 1,2,3,\ldots$$

where a is the width of the slit and λ is the wavelength of the light.

Problem:
The first minimum for light of wavelength 650 nm diffracted from a single slit is found at θ = 15 degrees. What is the width of the slit?

Solution:

Since m = 1 at the first minimum, $a = \dfrac{\lambda}{\sin\theta} = \dfrac{650}{\sin 15^0} = 2.5\mu m$.

Skill 24.5   Apply the superposition principle to determine characteristics of a resultant wave

According to the **principle of linear superposition**, when two or more waves exist in the same place, the resultant wave is the sum of all the waves, i.e. the amplitude of the resulting wave at a point in space is the sum of the amplitudes of each of the component waves at that point. Interference is usually observed in coherent waves, well-correlated waves that have very similar frequencies or even come from the same source.

Superposition of waves may result in either constructive or destructive interference. **Constructive interference** occurs when the crests of the two waves meet at the same point in time. Conversely, **destructive interference** occurs when the crest of one wave and the trough of the other meet at the same point in time. It follows, then, that constructive interference increases amplitude and destructive interference decreased it. We can also consider interference in terms of wave phase; waves that are out of phase with one another will interfere destructively while waves that are in phase with one another will interfere constructively. In the case of two simple sine waves with identical amplitudes, for instance, amplitude will double if the waves are exactly in phase and drop to zero if the waves are exactly 180° out of phase.

Additionally, **interference can create a standing wave**, a wave in which certain points always have amplitude of zero. Thus, the wave remains in a constant position. Standing waves typically results when two waves of the same frequency traveling in opposite directions through a single medium are superposed. View an animation of how interference can create a standing wave at the following URL:

http://www.glenbrook.k12.il.us/GBSSCI/PHYS/mmedia/waves/swf.html

All wavelengths in the EM spectrum can experience interference as discussed before. Similarly, we may be familiar with examples of interference in sound waves. When two sounds waves with slightly different frequencies interfere with each other, beat results. We hear a beat as a periodic variation in volume with a rate that depends on the difference between the two frequencies. You may have observed this phenomenon when listening to two instruments being tuned to match; beating will be heard as the two instruments approach the same note and disappear when they are perfectly in tune.

COMPETENCY 25.0   UNDERSTAND THE CHARACTERISTICS OF SOUND WAVES AND THE MEANS BY WHICH SOUND WAVES ARE PRODUCED AND TRANSMITTED

Skill 25.1   Explain the physical nature of sound waves (including intensity and intensity level)

Sound waves are mechanical waves that can travel through various kinds of media, solid, liquid and gas. They are **longitudinal waves** transmitted as variations in the pressure of the surrounding medium. These variations in pressure form waves that are termed sound waves or acoustic waves. When the particles of the medium are drawn close together it is called **compression**. When the particles of the medium are spread apart it is known as **rarefaction**.

Sound waves have different characteristics in different materials (see **Skill 17.2**). Boundaries between different media can result in partial or total reflection of sound waves. Thus, the phenomenon of echo takes place when a sound wave strikes a material of differing characteristics than the surrounding medium. The reflection of a voice off the walls of a room is a particular example.
The frequency of the harmonic acoustic wave, ω, determines the **pitch** of the sound in the same manner that the frequency determines the color of an electromagnetic harmonic wave. **A high pitch sound corresponds to a high frequency sound wave** and a low pitch sound corresponds to a low frequency sound wave. Similarly, the amplitude of the pressure determines the **loudness** or just as the amplitude of the electric (or magnetic) field **E** determines the brightness or intensity of a color. **The intensity of a sound wave is proportional to the square of its amplitude.**

The decibel scale is used to measure sound intensity. It originated in a unit known as the bel which is defined as the reduction in audio level over 1 mile of a telephone cable. Since the bel describes such a large variation in sound, it became more common to use the decibel, which is equal to 0.1 bel. A decibel value is related to the intensity of a sound by the following equation:

$$X_{dB} = 10 \log_{10}\left(\frac{X}{X_0}\right)$$

Where X_{dB} is the value of the sound in decibels
    X is the intensity of the sound
    X_0 is a reference value with the same units as X. X_0 is commonly taken to be the threshold of hearing at $10^{-12} W / m^2$.

It is important to note the logarithmic nature of the decibel scale and what this means for the relative intensity of sounds. The perception of the intensity of sound increases logarithmically, not linearly. Thus, an increase of 10 dB corresponds to an increase by one order of magnitude. For example, a sound that is 20 dB is not twice as loud as sound that is 10 dB; rather, it is 10 times as loud. A sound that is 30 dB will the 100 times as loud as the 10 dB sound.

Finally, let's equate the decibel scale with some familiar noises. Below are the decibel values of some common sounds.

Whispering voice: 20 dB
Quiet office: 60 dB
Traffic: 70 dB
Cheering football stadium: 110 dB
Jet engine (100 feet away): 150 dB
Space shuttle liftoff (100 feet away): 190 dB

Skill 25.2   Discuss factors that affect the speed of sound in different media

Mechanical waves rely on the local oscillation of each atom in a medium, but the material itself does not move; only the energy is transferred from atom to atom. Therefore the material through which the mechanical wave is traveling greatly affects the wave's propagation and speed. In particular, a material's elastic constant and density affect the speed at which a wave travels through it. Both of these properties of a medium can predict the extent to which the atoms will vibrate and pass along the energy of the wave. The general relationship between these properties and the speed of a wave in a solid is given by the following equation:

$$V = \sqrt{\frac{C_{ij}}{\rho}}$$

where V is the speed of sound, C_{ij} is the elastic constant or bulk modulus, and ρ is the material density. It is worth noting that the elastic constant differs depending on direction in anisotropic materials and the ij subscripts indicate that this directionality must be taken into account.

The speed of sound also varies with temperature since material characteristics are often dependent on temperature. At $0^0 C$, the speed of sound is 331m/s in air, 1402m/s in water and 6420m/s in Aluminum. Even though water is much denser than air, it is a lot more incompressible than air and its bulk modulus is larger than that of air by a larger factor. Therefore sound travels much faster in water than in air.

In the case of a stretched string, the velocity of a transverse wave passing through it is given by

$$v = \sqrt{\frac{\tau}{\mu}}$$

where τ is the tension in the string and μ is its linear density.

Skill 25.3 Solve problems involving resonance, harmonics, and overtones

Resonance, harmonics, and overtones can be used to discuss and explain a variety of physical phenomena, from the periodic change of a variable star to the oscillation of a compass needle due to the Earth's magnetic field. However, the simplest way to discuss 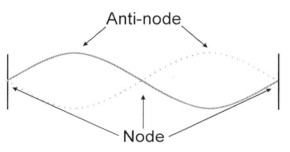 resonance, harmonics, and overtones is to use a standing wave, or a wave that is fixed at each end. The point in which the standing wave has no displacement is known as a *node* and *anti-nodes* are the points on the standing wave that have the greatest displacement. This is shown in the diagram above.

The **resonance** of a standing wave is the frequency at which the wave is vibrating. The resonate frequency of a standing wave can be found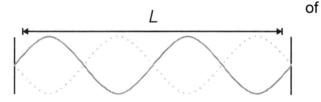

using the equation $\Rightarrow f = \dfrac{nv}{2L}$,

where *f* is the resonating frequency, *n* is the harmonic number, *v* is the speed of the wave, and *L* is the length of the standing wave between the two fixed points.

Problem: A string has a length of 2 m. When the string is plucked, the resulting wave has a velocity of 330 m/s and the harmonic number is 8 (n = 8). What is the resonating frequency?

Solution: Using the above formula$\Rightarrow f = \dfrac{nv}{2L}$ substitute in the given values $\Rightarrow$

$f = \dfrac{8 \times 330 m/s}{2 \times 2m} \Rightarrow f$ = 660Hz. The resonant frequency of the standing wave is 660 Hz.

The **harmonics** (n) of a standing wave are the natural resonant frequencies of the wave. The best way to visualize this is to use a length of string that is fixed at both ends. When plucked, the string will begin to vibrate forming a standing wave. The first harmonic (n=1) or the fundamental mode is when there is only one *anti-node* and the *nodes* are the two points at which the string is fixed *(Figure a)*. The second harmonic (n=2) or mode is when there are two anti-nodes and three nodes, as illustrated in *Figure b*. *Figures c* and *d* show the third and fourth harmonic, respectively. The first harmonic is the lowest natural frequency *(f)* that a standing wave can have. The second harmonic of a standing wave is the two times the frequency of the first harmonic (2*f*). The frequency of the third harmonic is equal to 3*f*. It can then be seen that the frequency of the nth harmonic is n*f*.

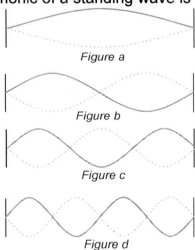

Figure a

Figure b

Figure c

Figure d

{Note: harmonics only exist as integer values i.e. n can only equal 1, 2, 3...}

Problem: The A string on an acoustic guitar has the lowest natural frequency of 440Hz. What are the frequencies of the 1st, 3rd, and 5th harmonics?

The first harmonic is $f \Rightarrow$ 440 Hz
The third harmonic is $3f \Rightarrow 3 \times 440$ = 1320 Hz
The fifth harmonic is $5f \Rightarrow 5 \times 440$ = 2200 Hz

The **overtone** of a standing wave is any natural resonant frequency or harmonic that is greater than the fundamental tone where the fundamental tone has the same frequency as the harmonic. This implies that the frequency of the second harmonic is equal to first the frequency of the first overtone. The table below shows the relationship between overtones and harmonics for the A string of a guitar.

Problem: A standing wave is oscillating with a frequency of 2640 Hz which is the 6th harmonic of the standing wave. What is the overtone number?

Solution: The frequency of 2640 Hz is the 5th overtone of the standing wave. From the table, it can be seen that the overtone number is one

Frequency (Hz)	Overtone number	Harmonic number (n)
440	Fundamental	1st
880	1st	2nd
1320	2nd	3rd
1760	3rd	4th

less than the harmonic number of the same standing wave.

COMPETENCY 26.0 UNDERSTAND THE PRODUCTION, PROPERTIES, AND CHARACTERISTICS OF ELECTROMAGNETIC WAVES

Skill 26.1 **The properties (e.g., energy, frequency, wavelength) of components (e.g., visible light, ultraviolet radiation) of the electromagnetic spectrum**

The electromagnetic spectrum is measured using frequency (f) in hertz or wavelength (λ) in meters. The frequency times the wavelength of every electromagnetic wave equals the speed of light (3.0×10^8 meters/second).

Roughly, the range of wavelengths of the electromagnetic spectrum is:

	$\underline{f}$	$\underline{\lambda}$
Radio waves	10^{5} - 10^{-1} hertz	10^{3} - 10^{9} meters
Microwaves	3×10^{9} - 3×10^{11} hertz	10^{-3} - 10^{-1} meters
Infrared radiation	3×10^{11} - 4×10^{14} hertz	7×10^{-7} - 10^{-3} meters
Visible light	4×10^{14} - 7.5×10^{14} hertz	4×10^{-7} - 7×10^{-7} meters
Ultraviolet radiation	7.5×10^{14} - 3×10^{16} hertz	10^{-8} - 4×10^{-7} meters
X-Rays	3×10^{16} - 3×10^{19} hertz	10^{-11} - 10^{-8} meters
Gamma Rays	$> 3 \times 10^{19}$ hertz	$< 10^{-11}$ meters

Radio waves are used for transmitting data. Common examples are television, cell phones, and wireless computer networks. Microwaves are used to heat food and deliver Wi-Fi service. Infrared waves are utilized in night vision goggles. Visible light we are all familiar with as the human eye is most sensitive to this wavelength range. Light of different colors have different wavelengths. In the visible range, red light has the largest wavelength while violet light has the smallest. UV light causes sunburns and would be even more harmful if most of it were not captured in the Earth's ozone layer. X-rays aid us in the medical field and gamma rays are most useful in the field of astronomy.

Skill 26.2   Give examples of applications of the components of the electromagnetic spectrum (e.g., infrared detectors, solar heating, x-ray machines)

The characteristics of electromagnetic radiation vary depending on the frequency of the radiation and the properties of the material through which it is propagating. Virtually innumerable applications have been devised for various frequency bands, leading to the development of a number of interesting and practical devices and inventions.

A common application of the microwave portion of the electromagnetic spectrum is the **microwave oven**. Since water molecules have a local charge imbalance, they each manifest a dipole moment. Microwave radiation operates at a wavelength such that incidence on water molecules imparts oscillating motion because of the dipole moments. This motion causes the temperature of the water to rise, thus cooking the food.

Another useful application of the electromagnetic spectrum is **remote communications.** A number of different frequency bands can be exploited, each with its own characteristics. Since oscillation of charges produces electromagnetic waves that propagate away from the source, antennas that are supplied with an electrical signal can be used to send or receive information through modulation of the waves. Two common modulation schemes are amplitude modulation (AM), for which the amplitude of a sine wave at a given frequency is varied, and frequency modulation (FM), for which the frequency of a sine wave (with a specific central frequency) is modulated. In both cases, these modulation schemes use a band of frequencies (bandwidth) in accordance with Fourier theory. Thus, a broadcast radio station is actually allocated a range of frequencies around a single central frequency (50 kHz around 100.5 MHz, for example). A virtually unlimited range of frequencies can be used for communication although some are more suited to certain situations than others. For example, line-of-sight communication is amenable to the use of a visible (or non-visible) laser as well as microwaves which are commonly used for cellular phones.

In cases where visible light is dim or absent, **infrared detectors or video equipment** can be used to produce images of scenes that appear dark to the unaided eye. Depending on their temperature, objects radiate infrared waves which can be used to produce false-color images of an otherwise dark environment.

Another method of imaging is the use of **X-rays**. These high-energy (i.e., high frequency) waves are able to penetrate certain materials, such as flesh, but are unable to penetrate other materials, such as bone or metal. Thus, X-rays may be used to visualize the skeletal structure of a human or animal without an invasive procedure or they may be used to detect hidden metallic items, such as various forms of contraband.

An advanced form of imaging is **holography**, which allows recording of a three-dimensional image, rather than just a two-dimensional image. Holography uses the coherent light produced by lasers to imprint an image on a photographic plate; in addition to amplitude information, which is all that is recorded in standard photography, holography also includes phase information, thus allowing capture of a three-dimensional image.

The broad range of frequencies radiated by the Sun can be collected in a number of ways to beneficially use or convert the incident energy. So-called **solar panels** can be used to convert electromagnetic radiation into electricity for powering devices or for storage in batteries. Also, the heat produced as a result of absorption of solar radiation can be used, for example, to heat a house. An appropriate system to collect the radiation, convert it to heat and distribute it throughout a house or other structure is one particular use of this phenomenon.

COMPETENCY 27.0 UNDERSTAND THE PRINCIPLES OF LENSES AND MIRRORS

Skill 27.1 Comparing types and characteristics of lenses and mirrors

Lenses

A lens is a device that causes electromagnetic radiation to converge or diverge. The most familiar lenses are made of glass or plastic and designed to concentrate or disperse visible light. Two of the most important parameters for a lens are its thickness and its focal length. Focal length is a measure of how strongly light is concentrated or dispersed by a lens. For a convex or converging lens, the focal length is the distance at which a beam of light will be focused to a single spot. Conversely, for a concave or diverging lens, the focal length is the distance to the point from which a beam appears to be diverging.

The images produced by lenses can be either virtual or real. A virtual image is one that is created by rays of light that appear to diverge from a certain point. Virtual images cannot be seen on a screen because the light rays do not actually meet at the point where the image is located. If an image and object appear on the same side of a converging lens, that image is defined as virtual. For virtual images, the image location will be negative and the magnification positive. Real images, on the other hand, are formed by light rays actually passing through the image. Thus, real images are visible on a screen. Real images created by a converging lens are inverted and have a positive image location and negative magnification.

Plane mirrors

Plane mirrors form virtual images. In other words, the image is formed behind the mirror where light does not actually reach. The image size is equal to the object size and object distance is equal to the image distance; i.e. the image is the same distance behind the mirror as the object is in front of the mirror. Another characteristic of plane mirrors is left-right reversal.

Example: Suppose you are standing in front of a mirror with your right hand raised. The image in the mirror will be raising its left hand.

Problem: If a cat creeps toward a mirror at a rate of 0.20 m/s, at what speed will the cat and the cat's image approach each other?

Solution: In one second, the cat will be 0.20 meters closer to the mirror. At the same time, the cat's image will be 0.20 meters closer to the cat. Therefore, the cat and its image are approaching each other at the speed of 0.40 m/s.

<u>Problem</u>: If an object that is two feet tall is placed in front of a plane mirror, how tall will the image of the object be?

<u>Solution</u>: The image of the object will have the same dimensions as the actual object, in this case, a height of two feet. This is because the magnification of an image in a plane mirror is 1.

Curved mirrors

Curved mirrors are usually sections of spheres. In a concave mirror the inside of the spherical surface is silvered while in a convex mirror it is the outside of the spherical surface that is silvered.

Terminology associated with spherical mirrors:

Principal axis: The line joining the center of the sphere (of which we imagine the mirror is a section) to the center of the reflecting surface.

Center of curvature: The center of the sphere of which the mirror is a section.

Vertex: The point on the mirror where the principal axis meets the mirror or the geometric center of the mirror.

Focal point: The point at which light rays traveling parallel to the principal axis will meet after reflection in a concave mirror. For a convex mirror, it is the point from which light rays traveling parallel to the principal axis will appear to diverge from after reflection. The focal point is midway between the center of curvature and the vertex.

Focal length: The distance between the focal point and the vertex.

Radius of curvature: The distance between the center of curvature and the vertex, i.e. the radius of the sphere of which the mirror is a section. The radius of curvature is twice the focal length.

Image characteristics for **concave mirrors**:
1)   If the object is located **beyond the center of curvature**, the image will be **real, inverted, smaller** and located between the focal point and center of curvature.
2)   If the object is located **at the center of curvature**, the image will be **real, inverted, of the same height** and also located at the center of curvature.
3)   If the object is located **between the center of curvature and focal point**, the image will be **real, inverted, larger** and located beyond the center of curvature.
4)   If the object is located **at the focal point no image is formed**.
5)   If the object is located **between the focal point and vertex**, the image will be **virtual, upright, larger** and located on the opposite side of the mirror.
6)   If the object is located **at infinity** (very far away), the image is **real, inverted, smaller** and located at the focal point.

For **convex mirrors**, the image is always **virtual, upright, reduced in size** and formed on the opposite side of the mirror.

Skill 27.2   Using a ray diagram to locate the focal point or point of image formation of a lens or mirror

Ray diagrams are a convenient way to visualize the propagation of waves and to perform reasonably accurate calculations of the effect of mirrors and lenses on light.

For mirrors, the focal point is either real (for concave mirrors) or virtual (for convex mirrors). In either case, the focal point is found by looking at the behavior of two parallel rays incident upon the mirror.

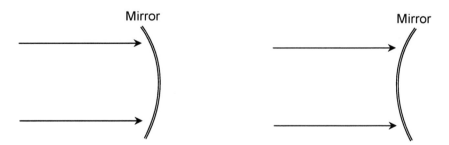

For each incident ray, the angle of reflection θ_r is equal to the angle of incidence θ_r, where the angle is measured from the normal to the mirror surface at the point of incidence.

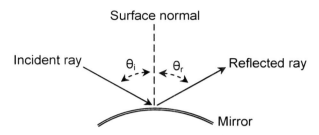

When this law is applied to the rays for either mirror, the focal points (f) are revealed as the (real or virtual) intersection of the rays. The virtual focus point is the intersection for the reflected rays when they are extended beyond the surface of the mirror.

The focal point of a lens is found in a similar manner, with the exception that instead of using the law of reflection, refraction by way of Snell's law must be applied. Since real lenses have a finite thickness, Snell's law must be used for the ray both as it enters the lens material and as it exits the lens material.

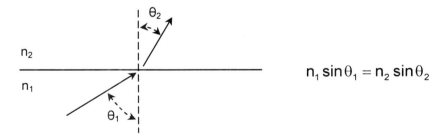

$$n_1 \sin\theta_1 = n_2 \sin\theta_2$$

As with reflection, refraction requires consideration of the normal to the surface. Also, the refractive indices must be used. It is assumed here that the refractive index of the outer medium is n_1 and the refractive index of the lens is n_2, and that $n_1 < n_2$. The intersection of parallel incident rays determines the focal point, which may again be either real or virtual. Real focal points occur on the opposite side of the lens from the source of illumination, and virtual focal points occur on the same side as the source of illumination.

Only one lens case is shown here, but the principles apply equally to all variations of lenses.

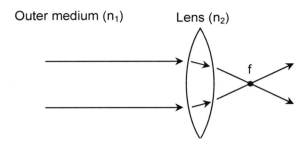

Care must be taken in properly identifying the normal and in applying Snell's law to both incidences of refraction for each ray.

Determining the point of image formation for mirrors and lenses is accomplished in a similar manner to that of the focal points. As with the focal points, image points may be either real or virtual, depending on the characteristics of the mirror or lens. To find the image point, two rays of differing angles must be traced as they interact with the lens or mirror. The initial directions of the rays can be chosen arbitrarily, but it is ideal to choose the directions such that the difficulty with determining the direction of the reflected or refracted ray is minimized. The intersection of these two rays is the image point. Only two examples are shown here, but the principles behind these examples may be applied to any variation of the situations, as well as to any combination of lenses and mirrors.

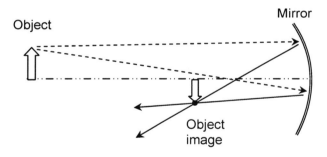

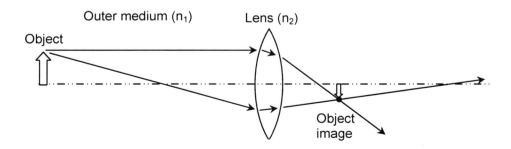

Skill 27.3 **Applying the lens and mirror equations to analyze problems involving lenses and mirrors**

Lenses

A thin lens in one in which focal length is much greater than lens thickness. For problems involving thin lenses, we can disregard any optical effects of the lens itself. Additionally, we can assume that the light that interacts with the lens makes a small angle with the optical axis of the system and so the sine and tangent values of the angle are approximately equal to the angle itself. This paraxial approximation, along with the thin lens assumptions, allows us to state:

$$\frac{1}{s} + \frac{1}{s'} = \frac{1}{f}$$

Where s=distance from the lens to the object (object location)
s'=distance from the lens to the image (image location)
f=focal length of the lens

Most lenses also cause some magnification of the object. Magnification is defined as:

$$m = \frac{y'}{y} = -\frac{s'}{s}$$

Where m=magnification
y'=image height
y=object height

Sign conventions will make it easier to understand thin lens problems:

Focal length: positive for a converging lens; negative for a diverging lens
Object location: positive when in front of the lens; negative when behind the lens
Image location: positive when behind the lens; negative when in front of the lens
Image height: positive when upright; negative when upside-down.
Magnification: positive for an erect, virtual image; negative for an inverted, real image

Problem:
A converging lens has a focal length of 10.00 cm and forms a 2.0 cm tall image of a 4.00 mm tall real object to the left of the lens. If the image is erect, is the image real or virtual? What are the locations of the object and the image?

Solution:
We begin by determining magnification:

$$m = \frac{y'}{y} = \frac{0.02m}{0.004m} = 5$$

Since the magnification is positive and the image is erect, we know the image must be virtual.

To find the locations of the object and image, we first relate them by using the magnification:

$$m = -\frac{s'}{s}$$

$$s' = -ms$$

Then we substitute into the thin lens equation, creating one variable in one unknown:

$$\frac{1}{s} + \frac{1}{s'} = \frac{1}{f}$$

$$\frac{1}{s} - \frac{1}{5s} = \frac{1}{10cm}$$

$$\frac{5-1}{5s} = \frac{1}{10cm}$$

$$s = \frac{40cm}{5} = 8cm \rightarrow \rightarrow s' = -5 \times 8cm = -40cm$$

Thus the object is located 8 cm to the left of the lens and the image is 40 cm to the left of the lens.

Mirrors

For image characteristics of curved mirrors see **Skill 19.1.** The relationship between the object distance from vertex s, the image distance from vertex s', and the focal length f is given by the equation

$$\frac{1}{s} + \frac{1}{s'} = \frac{1}{f}$$

Magnification is defined as:

$$m = \frac{y'}{y} = -\frac{s'}{s}$$

where m=magnification, y'=image height, y=object height

Problem: A concave mirror collects light from a star. If the light rays converge at 50 cm, what is the radius of curvature of the mirror?

Solution: The point at which the rays converge is known as the focal point. The focal length, in this case, 50 cm, is the distance from the focal point to the mirror. The radius of curvature is the distance from the vertex to the center of curvature. The vertex is the point on the mirror where the principal axis meets the mirror. The center of curvature represents the point in the center of the sphere from which the mirror was sliced. Since the focal point is the midpoint of the line from the vertex to the center of curvature, or focal length, the focal length would be one-half the radius of curvature. Since the focal length in this case is 50 cm, the radius of curvature would be 100 cm.

Problem: An image of an object in a mirror is upright and reduced in size. In what type of mirror is this image being viewed, plane, concave, or convex?

Solution: The image in a plane mirror would be the same size as the object. The image in a concave mirror would be magnified if upright. Only a convex mirror would produce a reduced upright image of an object.

Skill 27.4   Give examples of applications of lenses and mirrors (e.g., telescopes, compound microscopes, eyeglasses, etc.)

Simple magnifier

A simple magnifier is a convex or converging lens that allows a user to place an object closer to the eye than the near point (the distance within which objects become blurry, assumed to be approximately 25 cm) and view an enlarged virtual image. The magnification achieved is given by 25/f where f is the focal length of the magnifying lens.

Telescope

A telescope is a device that has the ability to make distant objects appear to be much closer. Most telescopes are one of two varieties, a refractor which uses lenses or a reflector which uses mirrors. Each accomplishes the same purpose but in totally different ways. The basic idea of a telescope is to collect as much light as possible, focus it, and then magnify it. The objective lens or primary mirror of a telescope brings the light from an object into focus. An eyepiece lens takes the focused light and "spreads it out" or magnifies it using the same principle as a magnifying glass using two curved surfaces to refract the light.

Microscope

Microscopes are used to view objects that are too small to be seen with the naked eye. A microscope usually has an objective lens that collects light from the sample and an eyepiece which brings the image into focus for the observer. It also has a light source to illuminate the sample. Typical optical microscopes achieve magnification of up to 1500 times.

Eye

The eye is a very complex sensory organ. Although there are many critical anatomical features of the eye, the lens and retina are the most important for focusing and sensing light. Light passes through the cornea and through the lens. The lens is attached to numerous muscles that contract and relax to move the lens in order to focus the light onto the retina. The pupil also contracts and relaxes to allow more or less light in the eye as required. The eye relies on refraction to focus light onto the retina. Refraction occurs at four curved interfaces; between the air and the front of the cornea, the back of the cornea and the aqueous humor, the aqueous humor at the front of the lens, and the back of the lens and the vitreous humor. When each of these interfaces are working properly the light arrives at the retina in perfect focus for transmission to the brain as an image.

Eye Glasses

When all the parts of the eye are not working together correctly, corrective lenses or eyeglasses may be needed to assist the eye in focusing the light onto the retina. The surfaces of the lens or cornea may not be smooth causing the light to refract in the wrong direction. This is called astigmatism. Another common problem is that the lens is not able to change its curvature appropriately to match the image. The cornea can also be misshaped resulting in blurred vision. Corrective lenses consist of curves pieces of glass which bend the light in order to change the focal point of the light. A nearsighted eye forms images in front of the retina. To correct this, a minus lens consisting of two concave prisms is used to bend light out and move the image back to the retina. A farsighted eye creates images behind the retina. This is corrected using plus lenses that bend light in and bring the image forward onto the retina. The worse the vision, the farther out of focus the image is on the retina. Therefore the stronger the lens the further the focal point is moved to compensate.

Spectroscope

Spectrometers known as spectroscopes are used to identify materials. Spectroscopes are used often in astronomy and some branches of chemistry. Early spectroscopes were simply a prism with graduations marking wavelengths of light. Modern spectroscopes typically use a diffraction grating, a movable slit, and some kind of photo detector, all automated and controlled by a computer. When materials are heated they emit light that is characteristic of its atomic composition. The emission of certain frequencies of light produce a pattern of lines that are comparable to a fingerprint. The yellow light emission of heated sodium is a typical example.

A spectroscope is able to detect, measure and record the frequencies of the emitted light. This is done by passing the light though a slit to a collimating lens which transforms the light into parallel rays. The light is then passed through a prism that refracts the beam into a spectrum of different wavelengths. The image is then viewed alongside a scale to determine the characteristic wavelengths. Spectral analysis is an important tool for determining and analyzing the composition of unknown materials as well as for astronomical studies.

Camera

A camera is another device that utilizes the lens' ability to refract light to capture and process an image. As with the eye, light enters the lens of a camera and focuses the light on the other side. Instead of focusing on the retina, the image is focused on the film to create a film negative. This film negative is later processed with chemicals to create a photograph. A camera uses a converging or convex lens. This lens captures and directs light to a single point to create a real image on the surface of the film. To focus a camera on an image, the distance of the lens from the film is adjusted in order to ensure that the real image converges on the surface of the film and not in front of or behind it.

Different lenses are available which capture and bend the light to different degrees. A lens with more pronounced curvature will be able to bend the light more acutely causing the image to converge more closely to the lens. Conversely a flatter lens will have a longer focal distance. The further the lens is located from the film (flatter lens), the larger the image becomes. Thus zoom lenses on cameras are flat while wide angle lenses are more rounded. The focal length number on a certain lens conveys the magnification ability of the lens.

The film functions like the retina of the eye in that it is light sensitive and can capture light images when exposed. However, this exposure must be brief to capture the contrasting amounts of light and a clear image. The rest of the camera functions to precisely control how much light contacts the film. The aperture is the lens opening which can open and close to let in more or less light. The temporal length of light exposure is controlled by the shutter which can be set at different speeds depending on the amount of action and level of light available. The film speed refers to the size of the light sensitive grains on the surface of the film. The larger grains absorb more light photons than the smaller grains, so film speed should be selected according to lighting conditions.

SUBAREA .V   **QUANTUM THEORY AND THE ATOM**

COMPETENCY 28.0   UNDERSTAND THE PRINCIPLES AND CONCEPTS OF THE PHOTOELECTRIC EFFECT, QUANTUM THEORY, AND THE DUAL NATURE OF LIGHT AND MATTER

Skill 28.1   Apply the laws of photoelectric emission to explain photoelectric phenomena

The wave theory of light explains many different phenomena but falls short when describing effects such as **blackbody radiation** and the **photoelectric effect**.

Blackbody radiation is the characteristic radiation of an ideal blackbody, i.e. a body that absorbs all the radiation incident upon it. Theoretical calculations of the frequency distribution of this radiation using classical physics showed that the energy density of this wave should increase as frequency increases. This result agreed with experiments at shorter wavelengths but failed at large wavelengths where experiment shows that that the energy density of the radiation actually falls back to zero.

In trying to resolve this impasse and derive the spectral distribution of blackbody radiation, Max Planck proposed that an atom can absorb or emit energy only in chunks known as quanta. The energy E contained in each quantum depends on the frequency of the radiation and is given by $E = hf$ where Planck's constant $h = 6.626 \times 10^{-34} J.s = 4.136 \times 10^{-15} eV.s$. Using this quantum hypothesis, Planck was able to provide an explanation for blackbody radiation that matched experiment.

Einstein extended Planck's idea further to suggest that quantization is a fundamental property of electromagnetic radiation which consists of quanta of energy known as **photons**. The energy of each photon is hf where h is Planck's constant.

Problem: A light beam has an intensity of 2W and wavelength of 600nm. What is the energy of each photon in the beam? How many photons are emitted by the beam every second?

Solution: The energy of each photon is given by
$E = hc / \lambda = 6.626 \times 10^{-34} \times 3 \times 10^8 / (600 \times 10^{-9}) = 3.31 \times 10^{-19} J$.
The number of photons emitted each second = $2 / (3.31 \times 10^{-19}) = 6.04 \times 10^{18}$.

Einstein used the photon hypothesis to explain the photoelectric effect.

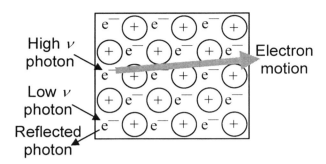

The **photoelectric effect** occurs when **light shining on a clean metal surface causes the surface to emit electrons**. The energy of an absorbed photon is transferred to an electron as shown to the right. If this energy is greater than the binding energy holding the electron close to nearby nuclei then the electron will move. A high energy (high frequency, low wavelength) photon will not only dislodge an electron from the "electron sea" of a metal but it will also impart kinetic energy to the electron, making it move rapidly. These electrons in motion will produce an electric current if a circuit is present.

When the metal surface on which light is incident is a cathode with the anode held at a higher potential V, an electric current flows in the external circuit. It is observed that current flows only for light of higher frequencies. Also there is a threshold negative potential, the **stopping potential** V_0 below which no current will flow in the circuit.

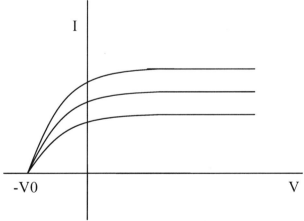

The figure displayed above shows current flow vs. potential for three different intensities of light. It shows that the maximum current flow increases with increasing light intensity but the stopping potential remains the same.

All these observations are counter-intuitive if one considers light to be a wave but may be understood in terms of light particles or photons. According to this interpretation, each photon transfers its energy to a single electron in the metal. Since the energy of a photon depends on its frequency, only a photon of higher frequency can transfer enough energy to an electron to enable it to pass the stopping potential threshold.

When V is negative, only electrons with a kinetic energy greater than $|eV|$ can reach the anode. The maximum kinetic energy of the emitted electrons is given by eV_0. This is expressed by Einstein's photoelectric equation as

$$(\tfrac{1}{2}mv^2)_{max} = eV_0 = hf - \varphi$$

where the **work function** φ is the energy needed to release an electron from the metal and is characteristic of the metal.

Problem: The work function for potassium is 2.20eV. What is the stopping potential for light of wavelength 400nm?

Solution:
$$eV_0 = hf - \varphi = hc/\lambda - \varphi = 4.136 \times 10^{-15} \times 3 \times 10^8 / (400 \times 10^{-9}) - 2.20 = 3.10 - 2.20 = 0.90eV$$

Thus stopping potential $V_0 = 0.90$V

Skill 28.2   Analyze bright-line spectra in terms of electron transitions

Quantum #	Radius
$n \to \infty$	$r_\infty \to \infty$
$\vdots$	$\vdots$
$n = 5$ ——	$r_5 = 25a_0$
$n = 4$ ——	$r_4 = 16a_0$
$n = 3$ ——	$r_3 = 9a_0$
$n = 2$ ——	$r_2 = 4a_0$
$n = 1$ ——	$r_1 = a_0$
$\oplus$ (H nucleus)	

An electron may exist at distinct radial distances (r_n) from the nucleus. These distances are proportional to the square of the **principal quantum number**, n. For a hydrogen atom (shown at left), the proportionality constant is called the **Bohr radius** ($a_0 = 5.29 \times 10^{-11}$ m). This value is the mean distance of an electron from the nucleus at the ground state of $n = 1$. The distances of other electron shells are found by the formula:
$$r_n = a_0 n^2 .$$
As $n \to \infty$, the electron is no longer part of the hydrogen atom. Ionization occurs and the atom become an H^+ ion.

A quantum of energy (ΔE) emitted from or absorbed by an electron transition is directly proportional to the frequency of radiation. The proportionality constant between them is **Planck's constant** ($h = 6.63 \times 10^{-34}$ J·s):

$$\Delta E = h\nu \quad \text{and} \quad \Delta E = \frac{hc}{\lambda} .$$

The energy of an electron (E_n) is inversely proportional to its radius from the nucleus. For a hydrogen atom, only the principle quantum number determines the energy of an electron by the **Rydberg constant** ($R_H = 2.18 \times 10^{-18}$ J):

$$E_n = -\frac{R_H}{n^2}.$$

The Rydberg constant is used to determine the energy of a photon emitted or absorbed by an electron transition from one shell to another in the H atom:

$$\Delta E = R_H \left(\frac{1}{n_{initial}^2} - \frac{1}{n_{final}^2} \right).$$

When a photon is absorbed, n_{final} is greater than $n_{initial}$, resulting in positive values corresponding to an endothermic process. Ionization occurs when sufficient energy is added for the atom to lose its electron from the ground state. This corresponds to an electron transition from $n_{initial} = 1$ to $n_{final} \rightarrow \infty$. The Rydberg constant is the energy required to ionize one atom of hydrogen. Photon emission causes negative values corresponding to an exothermic process because $n_{initial}$ is greater than n_{final}.

Planck's constant and the speed of light are often used to express the Rydberg constant in units of s^{-1} or length. The formulas below determine the photon frequency or wavelength corresponding to a given electron transition:

$$\nu_{photon} = \left(\frac{R_H}{h} \right) \left| \frac{1}{n_{initial}^2} - \frac{1}{n_{final}^2} \right| \quad \text{and} \quad \lambda_{photon} = \frac{1}{\left(\frac{R_H}{hc} \right) \left| \frac{1}{n_{initial}^2} - \frac{1}{n_{final}^2} \right|}.$$

These formulas **relate observed lines in the hydrogen spectrum to individual transitions** from one quantum state to another.

A simple optical spectroscope separates visible light into distinct wavelengths by passing the light through a prism or diffraction grating. When electrons in hydrogen gas are excited inside a discharge tube, the emission spectroscope shown below detects photons at four visible wavelengths.

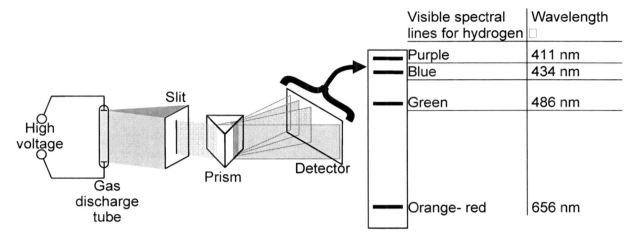

Visible spectral lines for hydrogen	Wavelength
Purple	411 nm
Blue	434 nm
Green	486 nm
Orange- red	656 nm

Every line in the hydrogen spectrum corresponds to a transition between electron energy levels. The four spectral lines from hydrogen emission spectroscopy in the visible range correspond to electron transitions from n = 3, 4, 5, and 6 to n =2 as shown in the table below.

Radiation type	Wavelength $\times$ (nm)	Frequency $\times$ (s^{-1})	Energy change $\times E$ (J)	Electron transition $n_{initial} \rightarrow n_{final}$
Ultraviolet	≤ 397	$\geq 7.55 \times 10^{14}$	$\leq -5.00 \times 10^{-19}$	$\infty \rightarrow 1, \ldots 2 \rightarrow 1$ $\infty \rightarrow 2, \ldots 7 \rightarrow 2$
Purple	411	7.31×10^{14}	-4.84×10^{-19}	$6 \rightarrow 2$
Blue	434	6.90×10^{14}	-4.58×10^{-19}	$5 \rightarrow 2$
Green	486	6.17×10^{14}	-4.09×10^{-19}	$4 \rightarrow 2$
Orange-red	656	4.57×10^{14}	-3.03×10^{-19}	$3 \rightarrow 2$
Infrared and beyond	≥ 821	$\leq 3.65 \times 10^{14}$	$\geq -2.42 \times 10^{-19}$	$\infty \rightarrow 3, \ldots 4 \rightarrow 3$ $\infty \rightarrow 4, \ldots 5 \rightarrow 4$ $\vdots$

Most lines in the hydrogen spectrum are not at visible wavelengths. Larger energy transitions produce ultraviolet radiation and smaller energy transitions produce infrared or longer wavelengths of radiation. Transitions between the first three and the first six energy levels of the hydrogen atom are shown in the diagram to the right. The energy transitions producing the four visible spectral lines are colored grey.

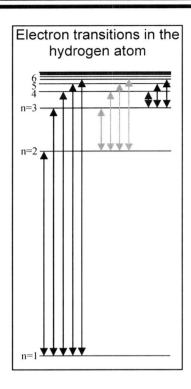

Electron transitions in the hydrogen atom

Skill 28.3   Applying the principles of stimulated emission of radiation to lasers and masers

The names "laser" and "maser" are acronyms for "light amplification by stimulated emission of radiation" and "microwave amplification by stimulated emission of radiation." Thus, as is evident by their names, these two devices are based on the principle of stimulated emission.

According to quantum theory, an atom in an excited state has a certain probability in a given time frame for relaxing to a lower state through, for example, the emission of a photon of the same energy as the energy difference between the two states. Stimulated emission, however, may take place when the excited atom is perturbed by a passing photon of the same energy as the excitation. In such a case, the atom relaxes to a lower energy by emitting a new photon, resulting in a total of two photons of equal frequency (and, thus, energy) and equal phase. That is to say, the photons are coherent.

In order to produce a significant level of stimulated emission for application in typical lasers and masers, population inversion is required. Population inversion is a situation in which there are more atoms in a particular excited state than in a particular lower-energy state. In order to achieve this condition, the material must be "pumped," which can be performed using electromagnetic fields (light). Pumping to produce population inversion often requires three or more atomic energy levels.

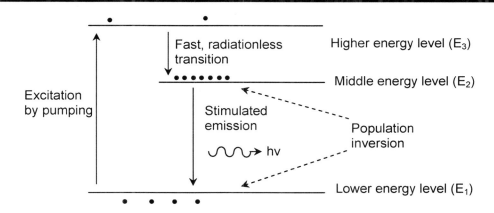

The above energy level diagram is for a so-called three-level laser (or maser). Other lasers may use more levels, depending on the material that is being pumped and the desired frequency output. The transition between the energy levels E_3 and E_2 is noted as being fast and radiationless; that is, almost as soon as the electron(s) of an atom are pumped to E_3, they decay to E_2 without radiating light. The energy difference can be emitted in the form of a phonon (vibration mode in the material, or heat).

The abundance of excited atoms due to population inversion allow for a "chain reaction" to form when photons of the proper frequency (proportional by Planck's constant to the energy difference between the excited state and lower-energy state) are incident. This results in a cascading increase in photon intensity through stimulated emission, thus producing the coherent, high-power light that is used by the laser or maser. This process has its limits, of course, as determined by how quickly the atoms can be pumped in relation to the relaxation processes. A saturation level exists beyond which the rate of stimulated emission cannot be increased. Spontaneously emitted photons can start the cascading process of stimulated emission.

The main components of a laser are a gain medium, an energy source (pump) for the gain medium and a resonant optical cavity with a partial transparency in one mirror. The resonant cavity is tuned to a particular frequency such that the photons of the desired laser frequency are coherent and, largely, isolated within the cavity. A partially transparent mirror on one end of the cavity allows some of the light to exit, thus producing the laser beam.

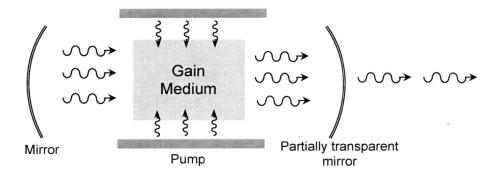

Masers, which operate at lower frequencies than lasers, generally rely on the same principles as lasers, although the types of gain media and resonant cavities may differ.

Skill 28.4   Analyzing evidence supporting the dual nature of light and matter

Scientists have argued for years whether light is a wave or a stream of particles. Actually, light exhibits the behaviors of both waves and particles.

Wave characteristics
Light undergoes **reflection**, **refraction**, and **diffraction** just as any wave would. The image you see in a mirrored surface is the result of the reflection of the light waves off the surface. Light waves follow the "law of reflection," i.e. the angle at which the light wave approaches a flat reflecting surface is equal to the angle at which it leaves the surface. When light crosses the boundary between two different media, its path is bent, or refracted. Diffraction occurs when light encounters an obstacle in its path or passes through an opening. Light diffracts around the sides of an object causing the shadow of the object to appear fuzzy.

Another phenomenon unique to waves is **wave interference**. This characteristic describes what happens when two waves meet while traveling along the same medium. If light **constructively interferes** (trough meets trough or crest meets crest), the two light waves reinforce one another to produce a stronger light wave. However, if light **destructively interferes** (crest meets trough), the two light waves destroy each other and no light wave is produced.

Polarization changes unpolarized light into polarized light. This process can only occur with a transverse wave. An everyday example of polarization is found in polarized sunglasses which reduce glare.

Particle characteristics

Einstein came up with the quantum theory of light that states that light is made up of **photons** or **quanta**, discrete particles of electromagnetic radiation. The photons, or individual particles of light, have been shown to have isolated arrival times. A movie was taken of the comet Hyakutake that showed a breakdown of the photons traveling with the comet and scattered throughout the region. Some phenomena such as **blackbody radiation** and the **photoelectric effect** can only be explained using the particle nature of light.

Presently, a combination of the two theories, or **wave - particle duality,** is accepted.

COMPETENCY 29.0 UNDERSTAND PHYSICAL MODELS OF ATOMIC STRUCTURE AND THE NATURE OF ELEMENTARY PARTICLES

Skill 29.1 Discuss historic and contemporary models of atomic structure (e.g., Bohr, Schrödinger, Heisenberg, Mayer, Bhabha)

In the West, the Greek philosophers Democritus and Leucippus first suggested the concept of the atom. They believed that all atoms were made of the same material but that varied sizes and shapes of atoms resulted in the varied properties of different materials. By the 19th century, John Dalton had advanced a theory stating that each element possesses atoms of a unique type. These atoms were also thought to be the smallest pieces of matter which could not be split or destroyed.

Atomic structure began to be better understood when, in 1897, JJ Thompson discovered the electron while working with cathode ray tubes. Thompson realized the negatively charged electrons were subatomic particles and formulated the "**plum pudding model**" of the atom to explain how the atom could still have a neutral charge overall. In this model, the negatively charged electrons were randomly present and free to move within a soup or cloud of positive charge. Thompson likened this to the dried fruit that is distributed within the English dessert plum pudding though the electrons were free to move in his model.

Ernest Rutherford disproved this model with the discovery of the nucleus in 1909. In **Rutherford's alpha scattering** experiments, he found that alpha particles striking a thin gold foil were scattered at large angles which indicated that the positive charge in an atom was concentrated in a small volume. Rutherford proposed a new "**planetary" model** of the atom in which electrons orbited around a positively charged nucleus like planets around the sun. Over the next 20 years, protons and neutrons (subnuclear particles) were discovered while additional experiments showed the inadequacy of the planetary model.

As quantum theory was developed and popularized (primarily by Max Planck and Albert Einstein), chemists and physicists began to consider how it might apply to atomic structure. Niels Bohr put forward a model of the atom in which electrons could only orbit the nucleus in circular orbitals with specific distances from the nucleus, energy levels, and angular momentums. In this model, electrons could only make instantaneous "quantum leaps" between the fixed energy levels of the various orbitals. The Bohr model of the atom was altered slightly by Arnold Sommerfeld in 1916 to reflect the fact that the orbitals were elliptical instead of round.

Though the Bohr model is still thought to be largely correct, it was discovered that electrons do not truly occupy neat, cleanly defined orbitals. Rather, they exist as more of an "electron cloud." The work of Louis de Broglie, Erwin Schrödinger, and Werner Heisenberg showed that an electron can actually be located at any distance from the nucleus. However, we can find the *probability* that the electrons exists at given energy levels (i.e., in particular orbitals) and those probabilities will show that the electrons are most frequently organized within the orbitals originally described in the Bohr model.

Skill 29.2   Demonstrate understanding of notation used to represent elements, molecules, ions, and isotopes

Every element in the periodic table is represented by its own symbol consisting of one or more letters. (E.g. H for hydrogen, He for Helium, O for oxygen). In general, a molecule of an element or a compound is represented by a formula that shows what atoms and how many of each constitute the molecule. For instance, water consists of two hydrogen and one oxygen atom. Thus a molecule of water is represented as H_2O. This is the kind of notation used in chemical equations.

Atoms and ions of a given element that differ in number of neutrons have a different mass and are called **isotopes**. In writing nuclear equations, where isotopes of the same element must be distinguished, it is helpful to use a notation where the number of nucleons and protons/electrons is included along with the element symbol. The identity of an **element** depends on the **number of protons** in the nucleus of the atom. This value is called the **atomic number** and it is sometimes written as a subscript before the symbol for the corresponding element. A nucleus with a specified number of protons and neutrons is called a **nuclide**, and a nuclear particle, either a proton or neutron, may be called a **nucleon**. The total number of nucleons is called the **mass number** and may be written as a superscript before the atomic symbol.

$$^{14}_{6}C$$ represents an atom of carbon with 6 protons and 8 neutrons.

The **number of neutrons** may be found by **subtracting the atomic number from the mass number**. For example, uranium- 235 has 235–92=143 neutrons because it has 235 nucleons and 92 protons.

An ion is an atom or molecule with a net positive or negative charge (due to an unequal number of protons and electrons) and is represented with a plus or minus sign and the number of excess electrons or protons placed at the top right-hand corner. For example, a positively charged sodium ion with one extra proton may be represented as Na^{+1}.

Skill 29.3 Discuss the relationship between the design of particle accelerators and elementary particle characteristics

Elementary (fundamental) particles are, in many ways, theoretical. These particles are not seen with the unaided (or aided) human eye, and therefore must be understood based on other phenomena that are, presumably, associated with or caused by their presence or action. As such, experimentally investigating fundamental particles requires a theoretical framework for design of the experimental apparatus and for interpretation of the results. With the recent advent of inexpensive computing power, complex and repetitive calculations can be conducted quickly and relatively inexpensively, but the numbers that are output by the computer are only as good as the data that is input and the interpretation that is later applied to the results.

First, it is critical to note that **the "resolution" of an experiment involving fundamental particles corresponds very closely to the kinetic energy of the particles involved**. As with light (photons), the wavelength of the illuminating particles must be smaller than the smallest desired dimension of resolution in order to "see" the object of interest. As such, a particle accelerator must be designed with the dimensions of the proposed investigation clearly in mind. That is to say, the energy capabilities of the accelerator must meet or exceed the requirements for a given resolution (such as, for example, investigation of quarks).

The method for accelerating the particles is key; fundamental particles, of course, cannot be accelerated reliably by mechanical or conventional means. The use of electric and magnetic fields and waves allows for the acceleration of particles, but only charged particles. Thus, if neutral particles are to be investigated in this manner, their production must be possible using charged particles (e.g., by way of charged particle collisions). The characteristics of the particles to be investigated using the accelerator must then be considered carefully and integrated into the design of the equipment.

Some form of target must be used to facilitate the production of collision events and to allow study of the particles. Using the analogy of light shining on an object, fundamental particles are used to "illuminate" an object (perhaps another fundamental particle) to examine its behavior and structure. The target may simply consist of another beam of particles traveling in the opposite direction, forcing the particles to collide. Alternatively, some fixed target may be used and may contain any number of materials, depending on the experiments of interest.

Last, **detection and transduction of data must be achieved through appropriate design of detectors and other equipment**. This equipment is responsible for converting an event associated with the detection of a particle into a signal recognizable by a computer and, ultimately, analyzable by the experimenters. Of interest in the design process is, for example, the charge of the particles that originate from a collision event. These particles can be deflected by electric or magnetic fields, and the associated detection equipment can take advantage of these properties. Photons, an example of an uncharged particle, can be detected with photomultiplier tubes. Thus, the types of detection equipment necessary in the design of an accelerator depend on the characteristics of the particles of interest.

The four main points in the design of an accelerator include the particle kinetic energy (and, concomitantly, the resolution) necessary for the proposed experiments, the form of particle accelerator appropriate to the particular particles, the target to be used for producing collision events and the detection equipment necessary for the specific particles being investigated. Of course, proper computer handling of the data is likewise crucial. Since computers simply output numbers, a proper theory for interpretation is necessary.

COMPETENCY 30.0 UNDERSTAND THE PRINCIPLES OF RADIOACTIVITY AND TYPES AND CHARACTERISTICS OF RADIATION, AND THE PROCESS OF RADIOACTIVE DECAY

Skill 30.1 Apply principles of the conservation of mass-energy and charge to balance equations for nuclear reactions

In alpha decay, an atom emits an alpha particle. An alpha particle contains two protons and two neutrons. This makes it identical to a helium nucleus and so an alpha particle may be written as He^{2+} or it can be denoted using the Greek letter α. Because a nucleus decaying through alpha radiation loses protons and neutrons, the mass of the atom loses about 4 Daltons and it actually becomes a different element (transmutation). For instance:

$$^{238}U \rightarrow {}^{234}Th + \alpha$$

This isotope of uranium weighs 238 Daltons. Because it is uranium, it has 92 protons, meaning it must have 146 neutrons.   When it undergoes alpha decay it loses 2 protons and 2 neutrons. The alpha particle weighs 4 Daltons and the nucleus that has undergone decay weighs 234 Daltons. Thus mass is conserved. The decayed nucleus will have a charge reduced by that of 2 protons following the decay. However, the emitted alpha particle carries the additional charge due to its 2 protons. Thus both charge and mass are conserved over all.

Like alpha decay, beta decay involves emission of a particle. In this case, though, it is a beta particle, which is either an electron or positron. Note that a positron is the antimatter equivalent of an electron and so these particles are often denoted β^- and β^+. Beta plus and minus decay occur via roughly opposite paths. In beta minus decay, a neutron is converted in a proton (specifically, a down quark is converted to an up quark), an electron and an anti-neutrino; the latter two are emitted. In beta plus decay, on the other hand, a proton is converted to a neutron, a positron, and a neutrino; again, the latter two are emitted. As in alpha decay, a nucleus undergoing beta decay is transmuted into a different element because the number of protons is altered. However, because the total number of nucleons remains unchanged, the atomic mass remains the same (note, that the neutron is actually slightly heavier than a proton so mass is gained during beta plus decay).

An example of beta minus decay is as follows:

$$_{55}^{137}Cs \rightarrow \, _{56}^{137}Ba + e^- + \bar{v}_e \qquad \text{(beta minus decay)}$$

This isotope of caesium weighs 137 Daltons and, like all caesium isotopes, it has 55 protons. When it undergoes beta minus decay, a neutron is converted to a proton and an electron and an anti-neutrino are lost. The total mass-energy of the system is conserved since the difference in mass between a neutron and an electron plus proton is balanced by the energy of the emitted electron and the anti-neutrino. In beta minus decay a neutron, with no charge, is split into a positively charged proton and a negatively charged electron. Thus the conservation of charge is satisfied. The electron is emitted, while the proton remains in the nucleus. With one extra proton, the nucleus is now a barium isotope.

In order to conserve mass-energy of the system, beta plus decay cannot occur in isolation but only in a nucleus since the mass of a neutron is greater than the mass of a proton plus electron. The difference in binding energy of the mother and daughter nucleus provides the additional energy needed for the reaction to go through. In the example below, charge is conserved when a positively charged proton is converted into a positively charged positron. With one fewer proton, the decayed nucleus is now a neon isotope weighing 22 Daltons.

$$_{11}^{22}Na \rightarrow \, _{10}^{22}Ne + e^+ + v_e \qquad \text{(beta plus decay)}$$

Problems involving balancing nuclear equations can involve simple radioactive decay, fission, fusion, and other nuclear processes. In all cases, both mass-energy and charge must be balanced.

Problem: Uranium 235 is used as a nuclear fuel in a chain reaction. The reaction is initiated by a single neutron and produces barium 141, an unknown isotope, and 3 neutrons that can go on to propagate the chain reaction. Determine the unknown isotope. Assume that the kinetic energies and the energy released in the reaction is negligible compared to the masses of the isotopes produced.

Solution: We can begin by writing out the reaction, leaving open the unknown isotope X.

$$_{92}^{235}U + \, _0^1 n \rightarrow \, _{56}^{141}Ba + X + 3\,_0^1 n$$

We begin with the charge balance; since the neutron has no charge, the unknown isotope must have 36 protons. Consulting a periodic table, we see that this will mean it is a Krypton isotope.

Now we can balance the mass. Since the original nucleus weighed 235 Daltons and one neutron was added to it, the total mass of the resultant nuclei must be 236. So, we can simple subtract the weight of the barium isotope and the 3 new neutrons to find the unknown weight:

$$236-141-3=92$$

Thus our unknown isotope is krypton 92, making the balanced equation:

$$^{235}_{92}U + ^{1}_{0}n \rightarrow ^{141}_{56}Ba + ^{92}_{36}Kr + 3^{1}_{0}n$$

Skill 30.2   Analyze radioactive decay in terms of the half-life concept

While the radioactive decay of an individual atom is impossible to predict, a mass of radioactive material will decay at a specific rate. Radioactive isotopes exhibit exponential decay and we can express this decay in a useful equation as follows:

$$A=A_0e^{kt}$$

Where A is the amount of radioactive material remaining after time t, A_0 is the original amount of radioactive material, t is the elapsed time, and k is the unique activity of the radioactive material. Note that k is unique to each radioactive isotope and it specifies how quickly the material decays. Sometimes it is convenient to express the rate of decay as half-life. **A half-life is the time needed for half a given mass of radioactive material to decay.** Thus, after one half-life, 50% of an original mass will have decayed, after two half lives, 75% will have decayed and so on.

Let's examine a sample problem related to radioactive decay.

Problem: Radiocarbon dating has been used extensively to determine the age of fossilized organic remains. It is based on the fact that while most of the carbon atoms in living things is ^{12}C, a small percentage is ^{14}C. Since ^{14}C is a radioactive isotope, it is lost from a fossilized specimen at a specific rate following the death of an organism. The original and current mass of ^{14}C can be inferred from the relative amount of ^{12}C. So, if the half-life of ^{14}C is 5730 years and a specimen that originally contained 1.28 mg of ^{14}C now contains 0.10 mg, how old is the specimen?

In certain problems, we may be simply provided with the activity, k, but in this problem we must use the information given about half-life to solve for k.

Since we know that after one half-life, 50% of the material remains radioactive, we can plug into the governing equation above:

$$A = A_0 e^{kt}$$

$$0.5\, A_0 = A_0 e^{5730k}$$

$$k = (\ln (0.5))/5730 = -0.0001209$$

Having determined k, we can use this same equation again to determine how old the specimen described above must be:

$$A = A_0 e^{kt}$$

$$0.10 = 1.28 e^{-0.0001209t}$$

$$t = \frac{\ln\left(\dfrac{0.10}{1.28}\right)}{-0.0001209} = 21087$$

Thus, the specimen is 21,087 years old.

Note that this same equation can be used to calculate the half-life of an isotope if information regarding the decay after a given number of years were provided.

Skill 30.3   Analyze the nuclear disintegration series for a given isotope

A nuclear disintegration series, or nuclear decay chain, is a description of the process of radioactive decay by which certain elemental isotopes transform into other elemental isotopes. The nuclear decay processes that are typically involved include alpha decay, which is emission of a helium (^{4}He) nucleus, and beta decay, which is emission of an electron (or positron) along with a corresponding neutrino. Beta decay can take the form of either a beta plus (β^+) decay or beta (β^-) minus decay. A β^+ decay involves proton decay into a neutron by way of emission of a positron and a neutrino; a β^- decay involves neutron decay into a proton by way of emission of an electron and an anti-neutrino. Thus, with a β^+ decay, an isotope reduces its atomic number by unity, but (approximately) maintains its atomic mass. In the case of a β^- decay, an isotope increases its atomic number by unity while, again, approximately maintaining its atomic mass. Some atomic mass difference actually exists due to the emission of energy during the decay. These decay processes are determined by an associated probability and are often quantified by a half-life, or the time required for half a sample to undergo a certain decay.

A particular isotope may go through a number of nuclear decays before reaching a stable form, depending on the characteristics of the weak interaction in the case of beta decays and on the strong interaction in the case of alpha decays. A decay chain (or nuclear disintegration) diagram can be used to depict the process of decay of a particular isotope as it approaches its stable form. Below is an example of such a diagram.

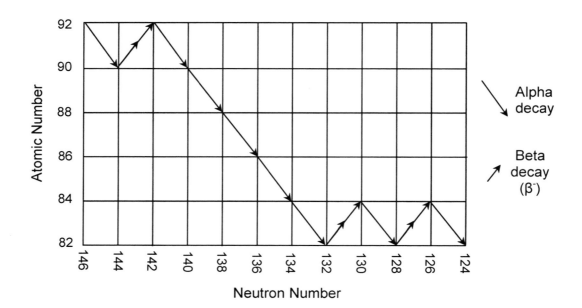

In the above diagram, the number of neutrons is plotted on the horizontal axis, and the atomic number (number of protons) is plotted on the vertical axis. The starting isotope (top, left) is uranium-238 (^{238}U), and the final, stable isotope (bottom, right) is lead-206 (^{206}Pb). In this decay chain, only alpha and β^- decays take place. Nevertheless, other chains may involve β^+ decays, which would be represented in the diagram as β^- decay arrows rotated by 180 degrees. As also can be seen in the above diagram, a nuclear disintegration chain may involve the same element in several instances, each as a different isotope. In the case of the ^{238}U chain, lead occurs as three different isotopes, with ^{206}Pb being the final, stable isotope. The half-lives of the above decays vary from billions of years to less than a second, depending on which particular decay is considered.

Skill 30.4   Basic operation of types of radiation detectors

Radiation detectors have been designed for a number of different applications and rely on various mechanisms for converting radiation into an electrical signal. Perhaps the most well-known form of radiation detector is the **Geiger counter**, a device based on the Geiger-Müller (GM) tube. The GM tube is an isolated chamber, usually filled with a combination of an inert gas, such as helium or neon, and a halogen gas, such as chlorine. Inside the tube, an anode and a cathode are present with a strong electric potential difference of hundreds of volts. When ionizing radiation enters the tube, some of the gas molecules are ionized with the positive ion being attracted towards the negatively charged anode and the negative ion (an electron) being attracted towards the positively charged cathode. The strong electric potential causes the ions to gain sufficient kinetic energy to further ionize other gas particles resulting in a cascade of particles and a detectable current. Depending on the specific materials used for the Geiger counter (e.g., the detection "window"), the device can detect alpha particles (helium nuclei), beta particles (an electron or positron resulting from a nuclear beta decay) or gamma rays (high-energy photons). So-called proportional counters, which operate on similar principles, are able to determine the energy of a particle of ionizing radiation through counting the "avalanches" that occur when gas molecules in the tube are ionized and accelerated to cause further ionization.

Photomultiplier tubes are an often-used method for detection of electromagnetic radiation in the visible region of the spectrum as well as in portions of the infrared and ultraviolet regions. The photoelectric effect is the driving phenomenon behind the operation of these devices. Photons that strike a so-called photocathode can cause the emission of an electron which is then accelerated towards a positively charged electrode called a dynode. A number of progressively more positively charged dynodes are arranged in the photomultiplier tube to cause a cascading effect. Each dynode produces further electrons in response to the impact of the incident electron, then "focuses" the emitted electrons onto the next dynode. Charge is collected at an anode, resulting in an electrical signal. The signal current produced by a photomultiplier tube is proportional to the intensity of the incident light.

Another device for detecting ionizing radiation is the **scintillation counter**. In this case a so-called scintillator is used. This is a material that produces a burst of light as a result of incident high-energy particles (such as photons). The incident high-energy particle loses some of its energy in the scintillator and the energy is converted into a number of lower-energy photons. A photomultiplier tube is then used to measure the intensity of the light pulse which yields a measure of the energy of the incident particle. Scintillation counters are used in such contexts as high-energy physics for detection equipment in particle accelerators.

Another radiation detector used in particle physics is the **calorimeter (or microcalorimeter)**. High-energy particles, rather than producing light, as with the scintillator, lose some of their energy to produce heat through incidence on the material used in the calorimeter. A thermometer is used to detect the change in the temperature of the material with the signal intensity providing information on the energy of the particle.

A number of other devices can be used in various contexts involving detection of radiation. Particle track detectors, such as bubble chambers or other devices containing solid, liquid or gaseous materials can be used to identify the passage of charged (or sometimes uncharged) particles. Trails left in the material due to incidence of various particles, which are influenced by a magnetic field, can be analyzed to determine the energy, charge and other features of the particles.

COMPETENCY 31.0   UNDERSTAND TYPES AND CHARACTERISTICS OF NUCLEAR REACTIONS, METHODS OF INITIATING AND CONTROLLING THEM, AND APPLICATIONS OF NUCLEAR REACTIONS TO THE GENERATION OF ELECTRICITY

Skill 31.1   Explain characteristics of fission and fusion reactions

Nuclear fusion involves the joining of several nuclei to form one heavier nucleus. Nuclear fission is the reverse of fusion, in that it is the splitting of a nucleus to form multiple lighter nuclei. Depending on the weight of the nuclei involved, both fission and fusion may result in either the absorption or release of energy. Iron and nickel have the largest binding energies per nucleon and so are the most stable nuclei. Thus, *fusion* releases energy when the two nuclei are lighter than iron or nickel and *fission* releases energy when the two nuclei are heavier than iron or nickel. Conversely, fusion will absorb energy when the nuclei are heavier and fission will absorb energy when the nuclei are lighter.

Nuclear fusion is common in nature and is the mechanism by which new natural elements are created. Fusion reactions power the stars and (energy absorbing) fusion of heavy elements occurs in supernova explosions. Despite the fact that significant energy is required to trigger the fusion of two nuclei (to overcome the electrostatic repulsion between the positively charged protons), the reaction can be self-sustaining because the energy released by the fusion of two light nuclei is greater than that required to force them together. Fusion is typically much harder to control that fission and so it is not used for power generation though fusion reactions are used to drive hydrogen bombs.

Nuclear fission is unique in that it can be harnessed for a variety of applications. This is done via the use of a chain reaction initiated by the bombardment of certain isotopes with free neutrons. When a nucleus is struck by a free neutron, it splits into smaller nuclei and also produces free neutrons, gamma rays, and alpha and beta particles. The free neutrons can then go onto interact with other nuclei and perpetuate the fission reaction. Isotopes, such as ^{235}U and ^{239}P that sustain the chain reaction are known as fissile and used for nuclear fuel. Because fission can be controlled via chain reaction, it is used in nuclear power generation. Uncontrolled fission reactions are also used in nuclear weapons, including the atomic bombs developed during the Manhattan Project and exploded over Hiroshima and Nagasaki in 1945.

Though it is currently in use in many locations, nuclear fission for power generation remains somewhat controversial. The amount of available energy per pound in nuclear fuel is millions of times that in fossil fuels. Additionally, nuclear power generation does not produce the air and water pollutants that are problematic byproducts of fossil fuel combustion. The currently used fission reactions, however, do produce radioactive waste that must be contained for thousands of years.

Skill 31.2 The operation of components of a nuclear reactor (e.g., moderator, fuel rods, control rods)

Nuclear fission involves the splitting of an atomic nucleus with a subsequent emission of energy in the form of photons with a specific frequency or other particles with a specific kinetic energy. As with combustion, the energy released from a fission reaction can be used to power turbines and to generate electricity. About 20% of the electrical power generated in the United States comes from nuclear reactors.

If controlled, the neutron flux resulting from spontaneously decaying radioactive materials can be used to instigate further decay. With Uranium (^{235}U, specifically), for example, a neutron from another decay can cause the atom to split, thus releasing several neutrons in addition to energy. These neutrons can also cause further decay, resulting in a **chain reaction**. Nuclear weapons involve an uncontrolled chain reaction that yields a tremendously powerful explosion; nuclear power generators take advantage of the same effect, but in a controlled manner.

In order to produce a chain reaction, rods filled with **nuclear fuel pellets** (such as ^{235}U) are used inside the reactor core. Since fission requires relatively slow-moving neutrons, a so-called **moderator** is needed to slow down the fast neutrons produced from fission elsewhere in the core, thus making the chain reaction possible. The moderator can be in the form of water (standard water or deuterium-based "heavy water"), which can also serve as a coolant. While these steps are necessary to sustain a nuclear reaction, it is also equally important to control the reaction. This is accomplished through the use of **control rods** which are filled with a neutron-absorbing material such as boron or silver. While these materials capture free neutrons, they are not themselves fissile and thus their presence limits the nuclear reaction in the core. The location of the control rods can be adjusted for dynamic control of the reaction. The entire apparatus of fuel rods, control rods and moderator (and coolant) is contained inside a pressure vessel.

The heat produced from the controlled nuclear reaction can be used in the same manner as heat from combustion to produce steam and turn a turbine, thus generating electricity. The hyperboloid towers typically seen at nuclear power plants serve the purpose of cooling.

Another issue revolving around the use of nuclear power is the ability to use a nuclear reactor to produce weapons-grade materials for use in explosive nuclear devices. This poses a tremendous threat since nuclear weapons are able to cause untold destruction as seen at the close of the Second World War in Japan. Furthermore, in the context of international terrorism, nuclear power plants in non-aggressive nations can still be a threat if attacked. Thus, the plants themselves, although they may be operating for only peaceful purposes, could be damaged in such a way that nuclear materials are released into the environment, or, in a worst-case scenario, a runaway nuclear reaction is initiated. Even otherwise benign human error can lead to catastrophic consequences. Thus, in light of these considerations, a debate continues concerning nuclear power as the dangers and the benefits are weighed.

For a discussion of nuclear waste see **Skill 31.4**.

Skill 31.3   Calculate nuclear mass defect and binding energy

Independently of one another, mass and energy are not necessarily conserved, as one may be converted to another in certain instances. Together, however, as mass-energy, they are indeed conserved. In most macroscopic circumstances, mass and energy are each conserved individually. On a microscopic scale, however, especially in the realm of particle physics, interconversion of mass and energy can take place to a significant extent. In these cases, the total conservation of mass-energy in a closed system is the most general conservation law.

Binding energy and nuclear mass defect
The binding energy and nuclear mass defect are two complementary concepts associated with the interaction of fundamental particles. The protons that form the nucleus of an atom, for example, are positively charged and, thus, have a mutual electrostatic repulsion. As a result, the energy of a system of closely packed protons is high. Nevertheless, the atomic nucleus is still quite stable. This stability results from a release of energy upon formation of the nucleus, thus lowering the total energy of the system. The released energy is called the binding energy. Conversely, for the reverse process, the binding energy is the energy required to split the nucleus into its component particles, and is often described generally as an energy per nucleon.

The source of the released binding energy during a fusion of protons (or protons and neutrons) into a nucleus is a portion of the mass of the system. According to special relativity, the mass of an object (whether rest or relativistic mass) has an equivalent energy ($E = mc^2$ in the case of rest mass). In the context of nuclear physics, protons that are fused into a nucleus are able to release energy at the expense of a portion of their mass. This nuclear mass defect is the emission of a certain portion of the mass of the constituent particles, in the form of energy, to lower the total energy of the whole (e.g., the nucleus). Quantum chromodynamics (QCD) provides a more fundamental explanation of this phenomenon through the strong nuclear force, or color force. QCD details the interactions of the quarks that compose the substructure of protons and neutrons by way of the color charge, which is a property of quarks, and gluons, which are the mediating boson for the strong force.

Conservation of mass-energy

In light of these basic concepts surrounding the conservation of mass-energy, the nuclear mass defect and binding energy can be calculated using the relativistic relationship of mass and energy. The calculation can be performed by noting the difference in mass between the constituent particles and the final product of a reaction.

<u>Problem</u>: What is the binding energy of a deuteron (a nucleus composed of a proton and a neutron)?

<u>Solution</u>: The binding energy is the equivalent energy resulting from the difference in mass between the deuteron and the constituent particles.

Proton mass = 1.6726×10^{-27} kg
Neutron mass = 1.6749×10^{-27} kg
Deuteron mass = 3.3436×10^{-27} kg

Mass difference: $\Delta m = \left(1.6726 \times 10^{-27} + 1.6749 \times 10^{-27}\right) kg - 3.3436 \times 10^{-27} kg$
$\Delta m = 3.9 \times 10^{-30} kg$

Binding energy: $\Delta m c^2 = 3.9 \times 10^{-30} kg \cdot \left(2.9979 \times 10^8 \, ^m\!/_s\right)^2 = 3.5051 \times 10^{-13} J$

Skill 31.4 **Give examples of the isotopes commonly used to fuel nuclear reactors and the problems associated with the waste products generated by nuclear reactions.**

Nuclear fuels consist mostly of heavy elements that can undergo fission. Common isotopes used in reactors include Uranium 235 and Plutonium 239. Light isotopes such as Tritium are used as fuel for nuclear fusion.

While nuclear power generation does not produce the same kind of carbon-based emissions as combustion of oil or coal, for example, there are other equally (if not more so) undesirable by-products. Waste products from the nuclear reaction are, themselves, often highly radioactive, posing an environmental threat. Ionizing radiation from such waste can cause tremendous damage to living organisms. Since there is no apparent way to easily neutralize these by-products, they must be stored securely to prevent contamination of the environment. This, of course, requires a storage facility and adequate isolation. Security and structural integrity of these facilities must be continually monitored.

Sample Test

DIRECTIONS: Read each item and select the best response.

1. **Which statement best describes a valid approach to testing a scientific hypothesis?**
    (Easy) (Skill 2.1)

    A. Use computer simulations to verify the hypothesis

    B. Perform a mathematical analysis of the hypothesis

    C. Design experiments to test the hypothesis

    D. All of the above

2. **Which description best describes the role of a scientific model of a physical phenomenon?**
    (Average Rigor) (Skill 2.1)

    A. An explanation that provides a reasonably accurate approximation of the phenomenon

    B. A theoretical explanation that describes exactly what is taking place

    C. A purely mathematical formulation of the phenomenon

    D. A predictive tool that has no interest in what is actually occurring

3. **Which situation calls might best be described as involving an ethical dilemma for a scientist?**
    (Rigorous) (Skill 2.2)

    A. Submission to a peer-review journal of a paper that refutes an established theory

    B. Synthesis of a new radioactive isotope of an element

    C. Use of a computer for modeling a newly-constructed nuclear reactor

    D. Use of a pen-and-paper approach to a difficult problem

4. **Which of the following is not a key purpose for the use of open communication about and peer-review of the results of scientific investigations?**
(Average Rigor) (Skill 2.4)

A. Testing, by other scientists, of the results of an investigation for the purpose of refuting any evidence contrary to an established theory

B. Testing, by other scientists, of the results of an investigation for the purpose of finding or eliminating any errors in reasoning or measurement

C. Maintaining an open, public process to better promote honesty and integrity in science

D. Provide a forum to help promote progress through mutual sharing and review of the results of investigations

5. **Which of the following aspects of the use of computers for collecting experimental data is not a concern for the scientist?**
(Rigorous) (Skill 3.1)

A. The relative speeds of the processor, peripheral, memory storage unit and any other components included in data acquisition equipment

B. The financial cost of the equipment, utilities and maintenance

C. Numerical error due to a lack of infinite precision in digital equipment

D. The order of complexity of data analysis algorithms

6. If a particular experimental observation contradicts a theory, what is the most appropriate approach that a physicist should take? *(Average Rigor) (Skill 3.3)*

   A. Immediately reject the theory and begin developing a new theory that better fits the observed results

   B. Report the experimental result in the literature without further ado

   C. Repeat the observations and check the experimental apparatus for any potential faulty components or human error, and then compare the results once more with the theory

   D. Immediately reject the observation as in error due to its conflict with theory

7. Which of the following is *not* an SI unit? *(Average Rigor) (Skill 4.1)*

   A. Joule

   B. Coulomb

   C. Newton

   D. Erg

8. Which of the following best describes the relationship of precision and accuracy in scientific measurements? *(Easy) (Skill 4.4)*

   A. Accuracy is how well a particular measurement agrees with the value of the actual parameter being measured; precision is how well a particular measurement agrees with the average of other measurements taken for the same value

   B. Precision is how well a particular measurement agrees with the value of the actual parameter being measured; accuracy is how well a particular measurement agrees with the average of other measurements taken for the same value

   C. Accuracy is the same as precision

   D. Accuracy is a measure of numerical error; precision is a measure of human error

9. **Which statement best describes a rationale for the use of statistical analysis in characterizing the numerical results of a scientific experiment or investigation?**
    (Average Rigor) (Skill 4.4)

    A. Experimental results need to be adjusted, through the use of statistics, to conform to theoretical predictions and computer models

    B. Since experiments are prone to a number of errors and uncertainties, statistical analysis provides a method for characterizing experimental measurements by accounting for or quantifying these undesirable effects

    C. Experiments are not able to provide any useful information, and statistical analysis is needed to impose a theoretical framework on the results

    D. Statistical analysis is needed to relate experimental measurements to computer-simulated values

10. **Which statement best characterizes the relationship of mathematics and experimentation in physics?**
    (Easy) (Skill 4.6)

    A. Experimentation has no bearing on the mathematical models that are developed for physical phenomena

    B. Mathematics is a tool that assists in the development of models for various physical phenomena as they are studied experimentally, with observations of the phenomena being a test of the validity of the mathematical model

    C. Mathematics is used to test the validity of experimental apparatus for physical measurements

    D. Mathematics is an abstract field with no relationship to concrete experimentation

11. Which of the following mathematical tools would not typically be used for the analysis of an electromagnetic phenomenon?
(Rigorous) (Skill 4.6)

A. Trigonometry

B. Vector calculus

C. Group theory

D. Numerical methods

12. For a problem that involves parameters that vary in rate with direction and location, which of the following mathematical tools would most likely be of greatest value?
(Rigorous) (Skill 4.8)

A. Trigonometry

B. Numerical analysis

C. Group theory

D. Vector calculus

13. Which of the following devices would be best suited for an experiment designed to measure alpha particle emissions from a sample?
(Average Rigor) (Skill 6.1)

A. Photomultiplier tube

B. Thermocouple

C. Geiger-Müller tube

D. Transistor

14. Which of the following experiments presents the most likely cause for concern about laboratory safety?
(Average Rigor) (Skill 6.2)

A. Computer simulation of a nuclear reactor

B. Vibration measurement with a laser

C. Measurement of fluorescent light intensity with a battery-powered photodiode circuit

D. Ambient indoor ionizing radiation measurement with a Geiger counter.

15. A brick and hammer fall from a ledge at the same time. They would be expected to:
(Easy) (Skill 7.2)

A. Reach the ground at the same time

B. Accelerate at different rates due to difference in weight

C. Accelerate at different rates due to difference in potential energy

D. Accelerate at different rates due to difference in kinetic energy

16. A baseball is thrown with an initial velocity of 30 m/s at an angle of 45°. Neglecting air resistance, how far away will the ball land?
(Rigorous) (Skill 7.2)

A. 92 m

B. 78 m

C. 65 m

D. 46 m

17. A skateboarder accelerates down a ramp, with constant acceleration of two meters per second squared, from rest. The distance in meters, covered after four seconds, is:
(Rigorous) (Skill 7.3)

A. 10

B. 16

C. 23

D. 37

18. When acceleration is plotted versus time, the area under the graph represents:
(Average Rigor) (Skill 7.4)

A. Time

B. Distance

C. Velocity

D. Acceleration

19. An inclined plane is tilted by gradually increasing the angle of elevation θ, until the block will slide down at a constant velocity. The coefficient of friction, μ$_k$, is given by:
(Rigorous) (Skill 8.1)

A. cos θ

B. sin θ

C. cosecant θ

D. tangent θ

20. An object traveling through air loses part of its energy of motion due to friction. Which statement best describes what has happened to this energy?
(Easy) (Skill 8.3)

A. The energy is destroyed

B. The energy is converted to static charge

C. The energy is radiated as electromagnetic waves

D. The energy is lost to heating of the air

21. The weight of an object on the earth's surface is designated *x*. When it is two earth's radii from the surface of the earth, its weight will be:
(Rigorous) (Skill 8.3)

A. *x*/4

B. *x*/9

C. 4*x*

D. 16*x*

22. Which of the following units is not used to measure torque?
(Average Rigor) (Skill 9.2)

A. slug ft

B. lb ft

C. N m

D. dyne cm

23. A uniform pole weighing 100 grams, that is one meter in length, is supported by a pivot at 40 centimeters from the left end. In order to maintain static position, a 200 gram mass must be placed _____ centimeters from the left end.
(Rigorous) (Skill 9.2)

A. 10

B. 45

C. 35

D. 50

24. The magnitude of a force is:
(Easy) (Skill 10.1)

A. Directly proportional to mass and inversely to acceleration

B. Inversely proportional to mass and directly to acceleration

C. Directly proportional to both mass and acceleration

D. Inversely proportional to both mass and acceleration

25. A projectile with a mass of 1.0 kg has a muzzle velocity of 1500.0 m/s when it is fired from a cannon with a mass of 500.0 kg. If the cannon slides on a frictionless track, it will recoil with a velocity of _____ m/s.
(Rigorous) (Skill 10.2)

A. 2.4

B. 3.0

C. 3.5

D. 1500

26. A car (mass m_1) is driving at velocity v, when it smashes into an unmoving car (mass m_2), locking bumpers. Both cars move together at the same velocity. The common velocity will be given by:
(Rigorous) (Skill 10.2)

A. m_1v/m_2

B. m_2v/m_1

C. $m_1v/(m_1 + m_2)$

D. $(m_1 + m_2)v/m_1$

27. A satellite is in a circular orbit above the earth. Which statement is false?
(Average Rigor) (Skill 11.1)

A. An external force causes the satellite to maintain orbit.

B. The satellite's inertia causes it to maintain orbit.

C. The satellite is accelerating toward the earth.

D. The satellite's velocity and acceleration are not in the same direction.

28. A 100 g mass revolving around a fixed point, on the end of a 0.5 meter string, circles once every 0.25 seconds. What is the magnitude of the centripetal acceleration?
(Average Rigor) (Skill 11.2)

A. 1.23 m/s^2

B. 31.6 m/s^2

C. 100 m/s^2

D. 316 m/s^2

29. Which statement best describes the relationship of simple harmonic motion to a simple pendulum of length L, mass m and displacement of arc length s?
(Average Rigor) (Skill 11.3)

A. A simple pendulum cannot be modeled using simple harmonic motion

B. A simple pendulum may be modeled using the same expression as Hooke's law for displacement s, but with a spring constant equal to the tension on the string

C. A simple pendulum may be modeled using the same expression as Hooke's law but with a spring constant equal to m g/L

D. A simple pendulum typically does not undergo simple harmonic motion

30. A mass of 2 kg connected to a spring undergoes simple harmonic motion at a frequency of 3 Hz. What is the spring constant?
(Average Rigor) (Skill 11.4)

A. 6 kg/s^2

B. 18 kg/s^2

C. 710 kg/s^2

D. 1000 kg/s^2

31. The kinetic energy of an object is _____ proportional to its _____.
(Average Rigor) (Skill 13.1)

A. Inversely...inertia

B. Inversely...velocity

C. Directly...mass

D. Directly...time

32. A force is given by the vector 5 N x + 3 N y (where x and y are the unit vectors for the x- and y- axes, respectively). This force is applied to move a 10 kg object 5 m, in the x direction. How much work was done?
(Rigorous) (Skill 13.1)

A. 250 J

B. 400 J

C. 40 J

D. 25 J

33. An office building entry ramp uses the principle of which simple machine?
(Easy) (Skill 13.3)

A. Lever

B. Pulley

C. Wedge

D. Inclined Plane

34. If the internal energy of a system remains constant, how much work is done by the system if 1 kJ of heat energy is added?
(Average Rigor) (Skill 16.1)

A. 0 kJ

B. -1 kJ

C. 1 kJ

D. 3.14 kJ

35. A calorie is the amount of heat energy that will:
(Easy) (Skill 16.2)

A. Raise the temperature of one gram of water from 14.5° C to 15.5° C.

B. Lower the temperature of one gram of water from 16.5° C to 15.5° C

C. Raise the temperature of one gram of water from 32° F to 33° F

D. Cause water to boil at two atmospheres of pressure.

36. An ice block at 0° Celsius is dropped into 100 g of liquid water at 18° Celsius. When thermal equilibrium is achieved, only liquid water at 0° Celsius is left. What was the mass, in grams, of the original block of ice?
Given:
1. Heat of fusion of ice = 80 cal/g
2. Heat of vaporization of ice = 540 cal/g
3. Specific Heat of ice = 0.50 cal/g°C
4. Specific Heat of water = 1 cal/g°C
(Rigorous) (Skill 16.2)

A. 2.0

B. 5.0

C. 10.0

D. 22.5

37. Heat transfer by electromagnetic waves is termed:
(Easy) (Skill 16.3)

A. Conduction

B. Convection

C. Radiation

D. Phase Change

38. A cooking thermometer in an oven works because the metals it is composed of have different:
(Average Rigor) (Skill 16.4)

A. Melting points

B. Heat convection

C. Magnetic fields

D. Coefficients of expansion

39. **Which of the following is not an assumption upon which the kinetic-molecular theory of gases is based?** *(Rigorous) (Skill 17.1)*

    A. Quantum mechanical effects may be neglected

    B. The particles of a gas may be treated statistically

    C. The particles of the gas are treated as very small masses

    D. Collisions between gas particles and container walls are inelastic

40. **What is temperature?** *(Average Rigor) (Skill 17.2)*

    A. Temperature is a measure of the conductivity of the atoms or molecules in a material

    B. Temperature is a measure of the kinetic energy of the atoms or molecules in a material

    C. Temperature is a measure of the relativistic mass of the atoms or molecules in a material

    D. Temperature is a measure of the angular momentum of electrons in a material

41. **Solids expand when heated because:** *(Rigorous) (Skill 17.2)*

    A. Molecular motion causes expansion

    B. $PV = nRT$

    C. Magnetic forces stretch the chemical bonds

    D. All material is effectively fluid

42. **What should be the behavior of an electroscope, which has been grounded in the presence of a positively charged object (1), after the ground connection is removed and then the charged object is removed from the vicinity (2)?** *(Average Rigor) (Skill 18.1)*

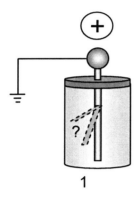

1   2

A. The metal leaf will start deflected (1) and then relax to an undeflected position (2)

B. The metal leaf will start in an undeflected position (1) and then be deflected (2)

C. The metal leaf will remain undeflected in both cases

D. The metal leaf will be deflected in both cases

43. **The electric force in Newtons, on two small objects (each charged to – 10 microCoulombs and separated by 2 meters) is:** *(Rigorous) (Skill 18.2)*

A. 1.0

B. 9.81

C. 31.0

D. 0.225

44. **A 10 ohm resistor and a 50 ohm resistor are connected in parallel. If the current in the 10 ohm resistor is 5 amperes, the current (in amperes) running through the 50 ohm resistor is:** *(Rigorous) (Skill 19.1)*

A. 1

B. 50

C. 25

D. 60

45. How much power is dissipated through the following resistive circuit? *(Average Rigor) (Skill 19.4)*

   A. 0 W

   B. 0.22 W

   C. 0.31 W

   D. 0.49 W

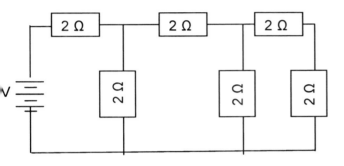

46. The greatest number of 100 watt lamps that can be connected in parallel with a 120 volt system without blowing a 5 amp fuse is: *(Rigorous) (Skill 19.4)*

   A. 24

   B. 12

   C. 6

   D. 1

47. Which of the following statements may be taken as a legitimate inference based upon the Maxwell equation that states $\nabla \cdot \mathbf{B} = 0$? *(Average Rigor) (Skill 20.1)*

   A. The electric and magnetic fields are decoupled

   B. The electric and magnetic fields are mediated by the W boson

   C. There are no photons

   D. There are no magnetic monopoles

48. What effect might an applied external magnetic field have on the magnetic domains of a ferromagnetic material? *(Rigorous) (Skill 20.1)*

   A. The domains that are not aligned with the external field increase in size, but those that are aligned decrease in size

   B. The domains that are not aligned with the external field decrease in size, but those that are aligned increase in size

   C. The domains align perpendicular to the external field

   D. There is no effect on the magnetic domains

49. **What is the effect of running current in the same direction along two parallel wires, as shown below?**
(Rigorous) (Skill 20.4)

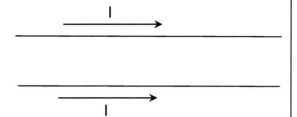

A. There is no effect

B. The wires attract one another

C. The wires repel one another

D. A torque is applied to both wires

50. **The current induced in a coil is defined by which of the following laws?**
(Easy) (Skill 21.1)

A. Lenz's Law

B. Burke's Law

C. The Law of Spontaneous Combustion

D. Snell's Law

51. **A light bulb is connected in series with a rotating coil within a magnetic field. The brightness of the light may be increased by any of the following except:**
(Average Rigor) (Skill 21.1)

A. Rotating the coil more rapidly.

B. Using more loops in the coil.

C. Using a different color wire for the coil.

D. Using a stronger magnetic field.

52. **What is the direction of the magnetic field at the center of the loop of current (I) shown below (i.e., at point A)?**
(Easy) (Skill 21.2)

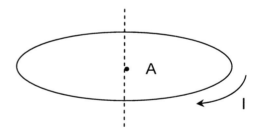

A. Down, along the axis (dotted line)
B. Up, along the axis (dotted line)
C. The magnetic field is oriented in a radial direction
D. There is no magnetic field at point A

53. The use of two circuits next to each other, with a change in current in the primary circuit, demonstrates:
(Rigorous) (Skill 21.4)

A. Mutual current induction

B. Dielectric constancy

C. Harmonic resonance

D. Resistance variation

54. A semi-conductor allows current to flow:
(Easy) (Skill 22.1)

A. Never

B. Always

C. As long as it stays below a maximum temperature

D. When a minimum voltage is applied

55. All of the following use semi-conductor technology, except a(n):
(Average Rigor) (Skill 22.1)

A. Transistor

B. Diode

C. Capacitor

D. Operational Amplifier

56. A wave generator is used to create a succession of waves. The rate of wave generation is one every 0.33 seconds. The period of these waves is:
(Average Rigor) (Skill 23.1)

A. 2.0 seconds

B. 1.0 seconds

C. 0.33 seconds

D. 3.0 seconds

57. **An electromagnetic wave propagates through a vacuum. Independent of its wavelength, it will move with constant:**
(Easy) (Skill 23.2)

   A. Acceleration

   B. Velocity

   C. Induction

   D. Sound

58. **A wave has speed 60 m/s and wavelength 30,000 m. What is the frequency of the wave?**
(Average Rigor) (Skill 23.2)

   A. 2.0×10^{-3} Hz

   B. 60 Hz

   C. 5.0×10^{2} Hz

   D. 1.8×10^{6} Hz

59. **Rainbows are created by:**
(Easy) (Skill 24.1)

   A. Reflection, dispersion, and recombination

   B. Reflection, resistance, and expansion

   C. Reflection, compression, and specific heat

   D. Reflection, refraction, and dispersion

60. **Which of the following is *not* a legitimate explanation for refraction of light rays at boundaries between different media?**
(Rigorous) (Skill 24.1)

   A. Light seeks the path of least time between two different points

   B. Due to phase matching and other boundary conditions, plane waves travel in different directions on either side of the boundary, depending on the material parameters

   C. The electric and magnetic fields become decoupled at the boundary

   D. Light rays obey Snell's law

61. A stationary sound source produces a wave of frequency *F*. An observer at position A is moving toward the horn, while an observer at position B is moving away from the horn. Which of the following is true?
(Rigorous) (Skill 24.2)

A. $F_A < F < F_B$

B. $F_B < F < F_A$

C. $F < F_A < F_B$

D. $F_B < F_A < F$

62. A monochromatic ray of light passes from air to a thick slab of glass (n = 1.41) at an angle of 45° from the normal. At what angle does it leave the air/glass interface?
(Rigorous) (Skill 24.3)

A. 45°

B. 30°

C. 15°

D. 55°

63. If one sound is ten decibels louder than another, the ratio of the intensity of the first to the second is:
(Average Rigor) (Skill 25.1)

A. 20:1

B. 10:1

C. 1:1

D. 1:10

64. The velocity of sound is greatest in:
(Average Rigor) (Skill 25.2)

A. Water

B. Steel

C. Alcohol

D. Air

65. A vibrating string's frequency is _____ proportional to the _____.
(Rigorous) (Skill 25.2)

A. Directly; Square root of the tension

B. Inversely; Length of the string

C. Inversely; Squared length of the string

D. Inversely; Force of the plectrum

66. Which of the following apparatus can be used to measure the wavelength of a sound produced by a tuning fork?
(Average Rigor) (Skill 25.3)

A. A glass cylinder, some water, and iron filings

B. A glass cylinder, a meter stick, and some water

C. A metronome and some ice water

D. A comb and some tissue

67. The highest energy is associated with:
(Easy) (Skill 26.1)

A. UV radiation

B. Yellow light

C. Infrared radiation

D. Gamma radiation

68. An object two meters tall is speeding toward a plane mirror at 10 m/s. What happens to the image as it nears the surface of the mirror?
(Rigorous) (Skill 27.1)

A. It becomes inverted.

B. The Doppler Effect must be considered.

C. It remains two meters tall.

D. It changes from a real image to a virtual image.

69. Automobile mirrors that have a sign, "objects are closer than they appear" say so because:
(Rigorous) (Skill 27.2)

A. The real image of an obstacle, through a converging lens, appears farther away than the object.

B. The real or virtual image of an obstacle, through a converging mirror, appears farther away than the object.

C. The real image of an obstacle, through a diverging lens, appears farther away than the object.

D. The virtual image of an obstacle, through a diverging mirror, appears farther away than the object.

70. If an object is 20 cm from a convex lens whose focal length is 10 cm, the image is:
(Rigorous) (Skill 27.2)

A. Virtual and upright

B. Real and inverted

C. Larger than the object

D. Smaller than the object

71. The constant of proportionality between the energy and the frequency of electromagnetic radiation is known as the:
(Easy) (Skill 28.1)

A. Rydberg constant

B. Energy constant

C. Planck constant

D. Einstein constant

72. Which phenomenon was first explained using the concept of quantization of energy, thus providing one of the key foundational principles for the later development of quantum theory?
(Rigorous) (Skill 28.1)

A. The photoelectric effect

B. Time dilation

C. Blackbody radiation

D. Magnetism

73. **Which statement best describes why population inversion is necessary for a laser to operate?**
(Rigorous) (Skill 28.3)

A. Population inversion prevents too many electrons from being excited into higher energy levels, thus preventing damage to the gain medium.

B. Population inversion maintains a sufficient number of electrons in a higher energy state so as to allow a significant amount of stimulated emission.

C. Population inversion prevents the laser from producing coherent light.

D. Population inversion is not necessary for the operation of most lasers.

74. **Bohr's theory of the atom was the first to quantize:**
(Average Rigor) (Skill 29.1)

A. Work

B. Angular Momentum

C. Torque

D. Duality

75. **Two neutral isotopes of a chemical element have the same numbers of:**
(Easy) (Skill 29.2)

A. Electrons and Neutrons

B. Electrons and Protons

C. Protons and Neutrons

D. Electrons, Neutrons, and Protons

76. **When a radioactive material emits an alpha particle only, its atomic number will:**
(Average Rigor) (Skill 30.1)

A. Decrease

B. Increase

C. Remain unchanged

D. Change randomly

77. **Ten grams of a sample of a radioactive material (half-life = 12 days) were stored for 48 days and re-weighed. The new mass of material was:**
(Rigorous) (Skill 30.2)

A. 1.25 g

B. 2.5 g

C. 0.83 g

D. 0.625 g

78. **Which of the following pairs of elements are not found to fuse in the centers of stars?**
(Average Rigor) (Skill 31.1)

   A. Oxygen and Helium

   B. Carbon and Hydrogen

   C. Beryllium and Helium

   D. Cobalt and Hydrogen

79. **In a fission reactor, heavy water:**
(Average Rigor) (Skill 31.2)

   A. Cools off neutrons to control temperature

   B. Moderates fission reactions

   C. Initiates the reaction chain

   D. Dissolves control rods

80. **Given the following values for the masses of a proton, a neutron and an alpha particle, what is the nuclear binding energy of an alpha particle?**
(Rigorous) (Skill 31.3)
Proton mass=1.6726×10^{-27} kg
Neutron mass=1.6749×10^{-27} kg
Alpha particle mass= 6.6465×10^{-27} kg

   A. 0 J

   B. 7.3417×10^{-27} J

   C. 4 J

   D. 4.3589×10^{-12} J

Free Response Questions

Question 1:

A 2 meter square loop of wire, connected to a resistive load, is arranged near a power line carrying 100 Amperes of current at 60 Hertz, as shown below. Assuming that the load is negligible in size, calculate the voltage amplitude across the load.

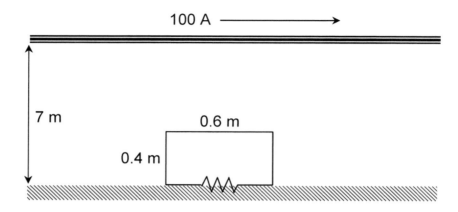

Question 2:

A 2,000 kg space vessel is designed to be accelerated away from the Sun using radiation pressure exerted on a sail of area 100 m^2 that has 95% reflectivity. (Assume that the sail absorbs the rest of the photons.) If the photon flux is assumed to be uniform over the space of interest, and if it has nominal average values of about 3.8×10^{21} photons m^{-2} s^{-1} and 5.5×10^{14} Hz frequency, what is the acceleration of the space vessel?

Answer Key

1.	D	25.	B	49.	B	73.	B
2.	A	26.	C	50.	A	74.	B
3.	B	27.	B	51.	C	75.	B
4.	A	28.	D	52.	A	76.	A
5.	D	29.	C	53.	A	77.	D
6.	C	30.	C	54.	D	78.	D
7.	D	31.	C	55.	C	79.	B
8.	A	32.	D	56.	C	80.	D
9.	B	33.	D	57.	B		
10.	B	34.	C	58.	A		
11.	C	35.	A	59.	D		
12.	D	36.	D	60.	C		
13.	C	37.	C	61.	B		
14.	B	38.	D	62.	B		
15.	A	39.	D	63.	B		
16.	A	40.	B	64.	B		
17.	B	41.	A	65.	A		
18.	C	42.	B	66.	B		
19.	D	43.	D	67.	D		
20.	D	44.	A	68.	C		
21.	B	45.	C	69.	D		
22.	A	46.	C	70.	B		
23.	C	47.	D	71.	C		
24.	C	48.	B	72.	C		

Rigor Analysis Table

Easy	21%	1,8,10,15,20,24,33,35,37,50,52,54,57,59,67,71,75
Average Rigor	39%	2,4,6,7,9,13,14,18,22,27,28,29,30,31,34,38,40,42, 45,47,51,55,56,58,63,64,66,74,76,78,79
Rigorous	40%	3,5,11,12,16,17,19,21,23,25,26,32,36,39,41,43,44, 46,48,49,53,60,61,62,65,68,69,70,72,73,77,80

Rationales with Sample Questions

1. **Which statement best describes a valid approach to testing a scientific hypothesis?**
    (Easy) (Skill 2.1)

    A. Use computer simulations to verify the hypothesis

    B. Perform a mathematical analysis of the hypothesis

    C. Design experiments to test the hypothesis

    D. All of the above

Answer: D

Each of the answers A, B and C can have a crucial part in testing a scientific hypothesis. Although experiments may hold more weight than mathematical or computer-based analysis, these latter two methods of analysis can be critical, especially when experimental design is highly time consuming or financially costly.

2. **Which description best describes the role of a scientific model of a physical phenomenon?**
    (Average Rigor) (Skill 2.1)

    A. An explanation that provides a reasonably accurate approximation of the phenomenon

    B. A theoretical explanation that describes exactly what is taking place

    C. A purely mathematical formulation of the phenomenon

    D. A predictive tool that has no interest in what is actually occurring

Answer: A

A scientific model seeks to provide the most fundamental and accurate description possible for physical phenomena, but, given the fact that natural science takes an *a posteriori* approach, models are always tentative and must be treated with some amount of skepticism. As a result, A is a better answer than B. Answers C and D overly emphasize one or another aspect of a model, rather than a synthesis of a number of aspects (such as a mathematical and predictive aspect).

3. **Which situation calls might best be described as involving an ethical dilemma for a scientist?**
    (Rigorous) (Skill 2.2)

    A. Submission to a peer-review journal of a paper that refutes an established theory

    B. Synthesis of a new radioactive isotope of an element

    C. Use of a computer for modeling a newly-constructed nuclear reactor

    D. Use of a pen-and-paper approach to a difficult problem

Answer: B

Although answer A may be controversial, it does not involve an inherently ethical dilemma, since there is nothing unethical about presenting new information if it is true or valid. Answer C, likewise, has no necessary ethical dimension, as is the case with D. Synthesis of radioactive material, however, involves an ethical dimension with regard to the potential impact of the new isotope on the health of others and on the environment. The potential usefulness of such an isotope in weapons development is another ethical consideration.

4. **Which of the following is not a key purpose for the use of open communication about and peer-review of the results of scientific investigations?**
    (Average Rigor) (Skill 2.4)

    A. Testing, by other scientists, of the results of an investigation for the purpose of refuting any evidence contrary to an established theory

    B. Testing, by other scientists, of the results of an investigation for the purpose of finding or eliminating any errors in reasoning or measurement

    C. Maintaining an open, public process to better promote honesty and integrity in science

    D. Provide a forum to help promote progress through mutual sharing and review of the results of investigations

Answer: A

Answers B, C and D all are important rationales for the use of open communication and peer-review in science. Answer A, however, would suggest that the purpose of these processes is to simply maintain the status quo; the history of science, however, suggests that this cannot and should not be the case.

5.   **Which of the following aspects of the use of computers for collecting experimental data is not a concern for the scientist?**
      (Rigorous) (Skill 3.1)

> A. The relative speeds of the processor, peripheral, memory storage unit and any other components included in data acquisition equipment
>
> B. The financial cost of the equipment, utilities and maintenance
>
> C. Numerical error due to a lack of infinite precision in digital equipment
>
> D. The order of complexity of data analysis algorithms

Answer: D

Although answer D might be a concern for later, when actual analysis of the data is undertaken, the collection of data typically does not suffer from this problem. The use of computers does, however, pose problems when, for example, a peripheral collects data at a rate faster than the computer can process it (A), or if the cost of running the equipment or of purchasing the equipment is prohibitive (B). Numerical error is always a concern with any digital data acquisition system, since the data that is collected is never exact.

6.   **If a particular experimental observation contradicts a theory, what is the most appropriate approach that a physicist should take?**
      (Average Rigor) (Skill 3.3)

      A. Immediately reject the theory and begin developing a new theory that better fits the observed results

      B. Report the experimental result in the literature without further ado

      C. Repeat the observations and check the experimental apparatus for any potential faulty components or human error, and then compare the results once more with the theory

      D. Immediately reject the observation as in error due to its conflict with theory

Answer: C

When experimental results contradict a reigning physical theory, as they do from time to time, it is almost never appropriate to immediately reject the theory (A) *or* the observational results (D). Also, since this is the case, reporting the result in the literature, without further analysis to provide an adequate explanation of the discrepancy, is unwise and unwarranted. Further testing is appropriate to determine whether the experiment is repeatable and whether any equipment or human errors have occurred. Only after further testing may the physicist begin to analyze the implications of the observational result.

7.   **Which of the following is *not* an SI unit?**
      (Average Rigor) (Skill 4.1)

      A. Joule

      B. Coulomb

      C. Newton

      D. Erg

Answer: D

The first three responses are the SI (*Le Système International d'Unités*) units for energy, charge and force, respectively. The fourth answer, the erg, is the CGS (centimeter-gram-second) unit of energy.

8. **Which of the following best describes the relationship of precision and accuracy in scientific measurements?**
   (Easy) (Skill 4.4)

   A. Accuracy is how well a particular measurement agrees with the value of the actual parameter being measured; precision is how well a particular measurement agrees with the average of other measurements taken for the same value

   B. Precision is how well a particular measurement agrees with the value of the actual parameter being measured; accuracy is how well a particular measurement agrees with the average of other measurements taken for the same value

   C. Accuracy is the same as precision

   D. Accuracy is a measure of numerical error; precision is a measure of human error

Answer: A

The accuracy of a measurement is how close the measurement is to the "true" value of the parameter being measured. Precision is how closely a group of measurements is to the mean value of all the measurements. By analogy, accuracy is how close a measurement is to the center of the bulls-eye, and precision is how tight a group is formed by multiple measurements, regardless of accuracy. Thus, measurements may be very precise and not very accurate, or they may be accurate but not overly precise, or they may be both or neither.

9. **Which statement best describes a rationale for the use of statistical analysis in characterizing the numerical results of a scientific experiment or investigation?**
    (Average Rigor) (Skill 4.4)

    A. Experimental results need to be adjusted, through the use of statistics, to conform to theoretical predictions and computer models

    B. Since experiments are prone to a number of errors and uncertainties, statistical analysis provides a method for characterizing experimental measurements by accounting for or quantifying these undesirable effects

    C. Experiments are not able to provide any useful information, and statistical analysis is needed to impose a theoretical framework on the results

    D. Statistical analysis is needed to relate experimental measurements to computer-simulated values

Answer: B

One of the main reasons for the use of statistical analysis is that various types of noise, errors and uncertainties can easily enter into experimental results. Among other things, statistics can help alleviate these difficulties by quantifying an average measurement value and a variance or standard deviation of the set of measurements. This helps determine the accuracy and precision of a set of experimental results. Answers A, C and D do not accurately describe ideal scientific experiments or the use of statistics.

10. **Which statement best characterizes the relationship of mathematics and experimentation in physics?**
    (Easy) (Skill 4.6)

    A. Experimentation has no bearing on the mathematical models that are developed for physical phenomena

    B. Mathematics is a tool that assists in the development of models for various physical phenomena as they are studied experimentally, with observations of the phenomena being a test of the validity of the mathematical model

    C. Mathematics is used to test the validity of experimental apparatus for physical measurements

    D. Mathematics is an abstract field with no relationship to concrete experimentation

Answer: B

Mathematics is used extensively in the study of physics for creating models of various phenomena. Since mathematics is abstract and not necessarily tied to physical reality, it must be tempered by experimental results. Although a particular theory may be mathematically elegant, it may have no explanatory power due to its inability to account for certain aspects of physical reality, or due to its inclusion of gratuitous aspects that seem to have no physical analog. Thus, experimentation is foundational, with mathematics being a tool for organizing and providing a greater context for observational results.

11. **Which of the following mathematical tools would not typically be used for the analysis of an electromagnetic phenomenon?**
    (Rigorous) (Skill 4.6)

    A. Trigonometry

    B. Vector calculus

    C. Group theory

    D. Numerical methods

Answer: C

Trigonometry and vector calculus are both key tools for solving problems in electromagnetics. These are, primarily, analytical methods, although they play a part in numerical analysis as well. Numerical methods are helpful for many problems that are otherwise intractable analytically. Group theory, although it may have some applications in certain highly specific areas, is generally not used in the study of electromagnetics.

12. **For a problem that involves parameters that vary in rate with direction and location, which of the following mathematical tools would most likely be of greatest value?**
    (Rigorous) (Skill 4.8)

    A. Trigonometry

    B. Numerical analysis

    C. Group theory

    D. Vector calculus

Answer: D

Each of the above answers might have some value for individual problems, but, generally speaking, those problems that deal with quantities that have direction and magnitude (vectors), and that deal with rates, would most likely be amenable to analysis using vector calculus (D).

13. **Which of the following devices would be best suited for an experiment designed to measure alpha particle emissions from a sample?**
    (Average Rigor) (Skill 6.1)

    A. Photomultiplier tube

    B. Thermocouple

    C. Geiger-Müller tube

    D. Transistor

Answer: C

The Geiger-Müller tube is the main component of the so-called Geiger counter, which is designed specifically for detecting ionizing radiation emissions, including alpha particles. The photomultiplier tube is better suited to measurement of electromagnetic radiation closer to the visible range (A), and the thermocouple is better suited to measurement of temperature (B). Transistors may be involved in instrumentation, but they are not sensors.

14. **Which of the following experiments presents the most likely cause for concern about laboratory safety?**
(Average Rigor) (Skill 6.2)

    A. Computer simulation of a nuclear reactor

    B. Vibration measurement with a laser

    C. Measurement of fluorescent light intensity with a battery-powered photodiode circuit

    D. Ambient indoor ionizing radiation measurement with a Geiger counter.

Answer: B

Assuming no profoundly foolish acts, the use of a computer for simulation (A), measurement with a battery-powered photodiode circuit (C) and ambient radiation measurement (D) pose no particular hazards. The use of a laser (B) must be approached with care, however, as unintentional reflections or a lack of sufficient protection can cause permanent eye damage.

15. **A brick and hammer fall from a ledge at the same time. They would be expected to:**
(Easy) (Skill 7.2)

    A. Reach the ground at the same time

    B. Accelerate at different rates due to difference in weight

    C. Accelerate at different rates due to difference in potential energy

    D. Accelerate at different rates due to difference in kinetic energy

Answer: A

This is a classic question about falling in a gravitational field. All objects are acted upon equally by gravity, so they should reach the ground at the same time. (In real life, air resistance can make a difference, but not at small heights for similarly shaped objects.) In any case, weight, potential energy, and kinetic energy do not affect gravitational acceleration. Thus, the only possible answer is (A).

16. A baseball is thrown with an initial velocity of 30 m/s at an angle of 45°. Neglecting air resistance, how far away will the ball land? *(Rigorous) (Skill 7.2)*

    A. 92 m

    B. 78 m

    C. 65 m

    D. 46 m

Answer: A

To answer this question, recall the equations for projectile motion:
$y = \frac{1}{2} a t^2 + v_{0y} t + y_0$
$x = v_{0x} t + x_0$
where x and y are horizontal and vertical position, respectively; t is time; a is acceleration due to gravity; v_{0x} and v_{0y} are initial horizontal and vertical velocity, respectively; x_0 and y_0 are initial horizontal and vertical position, respectively.
For our case:
x_0 and y_0 can be set to zero
both v_{0x} and v_{0y} are (using trigonometry) = $(\sqrt{2} / 2)$ 30 m/s
$a = -9.81$ m/s^2

We then use the vertical motion equation to find the time aloft (setting y equal to zero to find the solution for t):
$0 = \frac{1}{2} (-9.81$ m/s$^2) t^2 + (\sqrt{2} / 2)$ 30 m/s t
Then solving, we find:
t = 0 s (initial set-up) or t = 4.324 s (time to go up and down)

Using t = 4.324 s in the horizontal motion equation, we find:
$x = ((\sqrt{2} / 2)$ 30 m/s$)$ (4.324 s)
$x = 91.71$ m

This is consistent only with answer (A).

17.   A skateboarder accelerates down a ramp, with constant acceleration
        of two meters per second squared, from rest. The distance in
        meters, covered after four seconds, is:
        (Rigorous) (Skill 7.3)

                A. 10

                B. 16

                C. 23

                D. 37

Answer: B

To answer this question, recall the equation relating constant acceleration to
distance and time:
$x = \frac{1}{2} a t^2 + v_0 t + x_0$ where x is position; a is acceleration; t is time; v_0 and x_0 are
initial velocity and position (both zero in this case)

thus, to solve for x:
$x = \frac{1}{2} (2 \text{ m/s}^2) (4^2 \text{s}^2) + 0 + 0$
$x = 16$ m

This is consistent only with answer (B).

18.   When acceleration is plotted versus time, the area under the graph
        represents:
        (Average Rigor) (Skill 7.4)

                A. Time

                B. Distance

                C. Velocity

                D. Acceleration

Answer: C
The area under a graph will have units equal to the product of the units of the two
axes. (To visualize this, picture a graphed rectangle with its area equal to length
times width.)
Therefore, multiply units of acceleration by units of time:
$(\text{length/time}^2)(\text{time})$
This equals length/time, i.e. units of velocity.

19.   An inclined plane is tilted by gradually increasing the angle of elevation θ, until the block will slide down at a constant velocity. The coefficient of friction, μ_k, is given by:
(Rigorous) (Skill 8.1)

A. cos θ

B. sin θ

C. cosecant θ

D. tangent θ

Answer: D

When the block moves, its force upstream (due to friction) must equal its force downstream (due to gravity).

The friction force is given by
$F_f = \mu_k N$
where μ_k is the friction coefficient and N is the normal force.

Using similar triangles, the gravity force is given by
$F_g = mg \sin \theta$
and the normal force is given by
$N = mg \cos \theta$

When the block moves at constant velocity, it must have zero net force, so set equal the force of gravity and the force due to friction:
$F_f = F_g$
$\mu_k\, mg \cos \theta = mg \sin \theta$
$\mu_k = \tan \theta$

Answer (D) is the only appropriate choice in this case.

20.   An object traveling through air loses part of its energy of motion due to friction. Which statement best describes what has happened to this energy?
(Easy) (Skill 8.3)

    A. The energy is destroyed

    B. The energy is converted to static charge

    C. The energy is radiated as electromagnetic waves

    D. The energy is lost to heating of the air

Answer: D

Since energy must be conserved, the energy of motion of the object is converted, in part, to energy of motion of the molecules in the air (and, to some extent, in the object). This additional motion is equivalent to an increase in heat. Thus, friction is a loss of energy of motion through heating.

21.   The weight of an object on the earth's surface is designated *x*. When it is two earth's radii from the surface of the earth, its weight will be:
(Rigorous) (Skill 8.3)

    A. *x*/4

    B. *x*/9

    C. 4*x*

    D. 16*x*

Answer: B

To solve this problem, apply the universal Law of Gravitation to the object and Earth:

$F_{gravity} = (GM_1M_2)/R^2$

Because the force of gravity varies with the square of the radius between the objects, the force (or weight) on the object will be decreased by the square of the multiplication factor on the radius. Note that the object on Earth's surface is *already* at one radius from Earth's center. Thus, when it is two radii from Earth's surface, it is three radii from Earth's center. R^2 is then nine, so the weight is *x*/9. Only answer (B) matches these calculations.

22.   **Which of the following units is not used to measure torque?**
       (Average Rigor) (Skill 9.2)

        A. slug ft

        B. lb ft

        C. N m

        D. dyne cm

Answer: A

To answer this question, recall that torque is always calculated by multiplying units of force by units of distance. Therefore, answer (A), which is the product of units of mass and units of distance, must be the choice of incorrect units. Indeed, the other three answers all could measure torque, since they are of the correct form. It is a good idea to review "English Units" before the teacher test, because they are occasionally used in problems.

23. **A uniform pole weighing 100 grams, that is one meter in length, is supported by a pivot at 40 centimeters from the left end. In order to maintain static position, a 200 gram mass must be placed _____ centimeters from the left end.**
(Rigorous) (Skill 9.2)

    A. 10

    B. 45

    C. 35

    D. 50

Answer: C

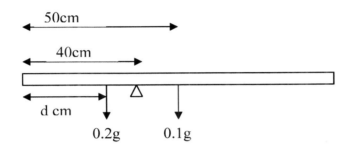

Since the pole is uniform, we can assume that its weight 0.1g acts at the center, i.e. 50 cm from the left end. In order to keep the pole balanced on the pivot, the 200 gram mass must be placed such that the torque on the pole due to the mass is equal and opposite to the torque due to the pole's weight. Thus, if the 200 gram mass is placed d cm from the left end of the pole,

$(40 - d) \times 0.2g = 10 \times 0.1g;\quad 40 - d = 5;\quad d = 35$ cm

24.   The magnitude of a force is:
       (Easy) (Skill 10.1)

       A. Directly proportional to mass and inversely to acceleration

       B. Inversely proportional to mass and directly to acceleration

       C. Directly proportional to both mass and acceleration

       D. Inversely proportional to both mass and acceleration

Answer: C

To solve this problem, recall Newton's 2nd Law, i.e. net force is equal to mass times acceleration. Therefore, the only possible answer is (C).

25.   **A projectile with a mass of 1.0 kg has a muzzle velocity of 1500.0 m/s when it is fired from a cannon with a mass of 500.0 kg. If the cannon slides on a frictionless track, it will recoil with a velocity of _____ m/s.**
       (Rigorous) (Skill 10.2)

       A. 2.4

       B. 3.0

       C. 3.5

       D. 1500

Answer: B

To solve this problem, apply Conservation of Momentum to the cannon-projectile system. The system is initially at rest, with total momentum of 0 kg m/s. Since the cannon slides on a frictionless track, we can assume that the net momentum stays the same for the system. Therefore, the momentum forward (of the projectile) must equal the momentum backward (of the cannon). Thus:

$p_{projectile} = p_{cannon}$
$m_{projectile} \, v_{projectile} = m_{cannon} \, v_{cannon}$
(1.0 kg)(1500.0 m/s) = (500.0 kg)(x)
x = 3.0 m/s
Only answer (B) matches these calculations.

26. A car (mass m_1) is driving at velocity v, when it smashes into an unmoving car (mass m_2), locking bumpers. Both cars move together at the same velocity. The common velocity will be given by: *(Rigorous) (Skill 10.2)*

   A. m_1v/m_2

   B. m_2v/m_1

   C. $m_1v/(m_1 + m_2)$

   D. $(m_1 + m_2)v/m_1$

Answer: C

In this problem, there is an inelastic collision, so the best method is to assume that momentum is conserved. (Recall that momentum is equal to the product of mass and velocity.)
Therefore, apply Conservation of Momentum to the two-car system:
Momentum at Start = Momentum at End
(Mom. of Car 1) + (Mom. of Car 2) = (Mom. of 2 Cars Coupled)
$m_1v + 0 = (m_1 + m_2)x$
$x = m_1v/(m_1 + m_2)$
Only answer (C) matches these calculations.

Watch out for the other answers, because errors in algebra could lead to a match with incorrect answer (D), and assumption of an elastic collision could lead to a match with incorrect answer (A).

27.   A satellite is in a circular orbit above the earth. Which statement is false?
(Average Rigor) (Skill 11.1)

   A. An external force causes the satellite to maintain orbit.

   B. The satellite's inertia causes it to maintain orbit.

   C. The satellite is accelerating toward the earth.

   D. The satellite's velocity and acceleration are not in the same direction.

Answer: B

To answer this question, recall that in circular motion, an object's inertia tends to keep it moving straight (tangent to the orbit), so a centripetal force (leading to centripetal acceleration) must be applied. In this case, the centripetal force is gravity due to the earth, which keeps the object in motion. Thus, (A), (C), and (D) are true, and (B) is the only false statement.

28.   A 100 g mass revolving around a fixed point, on the end of a 0.5 meter string, circles once every 0.25 seconds. What is the magnitude of the centripetal acceleration?
(Average Rigor) (Skill 11.2)

   A. 1.23 m/s^2

   B. 31.6 m/s^2

   C. 100 m/s^2

   D. 316 m/s^2

Answer: D

The centripetal acceleration is equal to the product of the radius and the square of the angular frequency ω. In this case, ω is equal to 25.1 Hz. Squaring this value and multiplying by 0.5 m yields the result in answer D.

29. **Which statement best describes the relationship of simple harmonic motion to a simple pendulum of length L, mass m and displacement of arc length s?**
   (Average Rigor) (Skill 11.3)

   A. A simple pendulum cannot be modeled using simple harmonic motion

   B. A simple pendulum may be modeled using the same expression as Hooke's law for displacement s, but with a spring constant equal to the tension on the string

   C. A simple pendulum may be modeled using the same expression as Hooke's law for displacement s, but with a spring constant equal to m g/L

   D. A simple pendulum typically does not undergo simple harmonic motion

Answer: C

The force on a simple pendulum may be expressed approximately (when displacement s is small) according to the following equation:

$$F \approx -\frac{mg}{L}s$$

This expression has the same form as Hooke's law (F = -kx). Thus, answer C is the most correct response. Another approach to the question is to eliminate answers A and D as obviously incorrect, and then to eliminate answer B as not having appropriate units for the spring constant.

30. A mass of 2 kg connected to a spring undergoes simple harmonic motion at a frequency of 3 Hz. What is the spring constant? *(Average Rigor) (Skill 11.4)*

    A. 6 kg/s^2

    B. 18 kg/s^2

    C. 710 kg/s^2

    D. 1000 kg/s^2

Answer: C

The spring constant, k, is equal to $m\omega^2$. In this case, ω is equal to 2π times the frequency of 3 Hz. The spring constant may be derived quickly by recognizing that the position of the mass varies sinusoidally with time at an angular frequency ω. Noting that the acceleration is the second derivative of the position with respect to time, the expression for k in Hooke's law (F = -kx) can be easily derived.

31. The kinetic energy of an object is _____ proportional to its _____.
    (Average Rigor) (Skill 13.1)

    A. Inversely…inertia

    B. Inversely…velocity

    C. Directly…mass

    D. Directly…time

Answer: C

To answer this question, recall that kinetic energy is equal to one-half of the product of an object's mass and the square of its velocity:
KE = ½ m v^2

Therefore, kinetic energy is directly proportional to mass, and the answer is (C). Note that although kinetic energy is associated with both velocity and momentum (a measure of inertia), it is not *inversely* proportional to either one.

32.   A force is given by the vector 5 N x + 3 N y (where x and y are the
       unit vectors for the x- and y- axes, respectively). This force is
       applied to move a 10 kg object 5 m, in the x direction. How much
       work was done?
       (Rigorous) (Skill 13.1)

           A. 250 J

           B. 400 J

           C. 40 J

           D. 25 J

Answer: D

To find out how much work was done, note that work counts only the force in the
direction of motion. Therefore, the only part of the vector that we use is the 5 N
in the x-direction. Note, too, that the mass of the object is not relevant in this
problem. We use the work equation:
Work = (Force in direction of motion) (Distance moved)
Work = (5 N) (5 m)
Work = 25 J
This is consistent only with answer (D).

33.   An office building entry ramp uses the principle of which simple
       machine?
       (Easy) (Skill 13.3)

           A. Lever

           B. Pulley

           C. Wedge

           D. Inclined Plane

Answer: D

To answer this question, recall the definitions of the various simple machines. A
ramp, which trades a longer traversed distance for a shallower slope, is an
example of an Inclined Plane, consistent with answer (D). Levers and Pulleys
act to change size and/or direction of an input force, which is not relevant here.
Wedges apply the same force over a smaller area, increasing pressure—again,
not relevant in this case.

34. If the internal energy of a system remains constant, how much work is done by the system if 1 kJ of heat energy is added?
(Average Rigor) (Skill 16.1)

A. 0 kJ

B. -1 kJ

C. 1 kJ

D. 3.14 kJ

Answer: C

According to the first law of thermodynamics, if the internal energy of a system remains constant, then any heat energy added to the system must be balanced by the system performing work on its surroundings. In the case of an ideal gas, the gas would necessarily expand when heated, assuming a constant internal energy was somehow maintained. Applying conservation of energy, answer C is found to be correct.

35. A calorie is the amount of heat energy that will:
(Easy) (Skill 16.2)

A. Raise the temperature of one gram of water from 14.5° C to 15.5° C.

B. Lower the temperature of one gram of water from 16.5° C to 15.5° C

C. Raise the temperature of one gram of water from 32° F to 33° F

D. Cause water to boil at two atmospheres of pressure.

Answer: A

The definition of a calorie is, "the amount of energy to raise one gram of water by one degree Celsius," and so answer (A) is correct. Do not get confused by the fact that 14.5° C seems like a random number. Also, note that answer (C) tries to confuse you with degrees Fahrenheit, which are irrelevant to this problem.

36. Use the information on heats below to solve this problem. An ice block at 0° Celsius is dropped into 100 g of liquid water at 18° Celsius. When thermal equilibrium is achieved, only liquid water at 0° Celsius is left. What was the mass, in grams, of the original block of ice?

Given: Heat of fusion of ice = 80 cal/g
Heat of vaporization of ice = 540 cal/g
Specific Heat of ice = 0.50 cal/g°C
Specific Heat of water = 1 cal/g°C
(Rigorous) (Skill 16.2)

A. 2.0

B. 5.0

C. 10.0

D. 22.5

Answer: D

To solve this problem, apply Conservation of Energy to the ice-water system. Any gain of heat to the melting ice must be balanced by loss of heat in the liquid water. Use the two equations relating temperature, mass, and energy:
$Q = m C \Delta T$ (for heat loss/gain from change in temperature)
$Q = m L$ (for heat loss/gain from phase change)
where Q is heat change; m is mass; C is specific heat; ΔT is change in temperature; L is heat of phase change (in this case, melting, also known as "fusion").

Then
$Q_{ice\ to\ water} = Q_{water\ to\ ice}$
(Note that the ice only melts; it stays at 0° Celsius—otherwise, we would have to include a term for warming the ice as well. Also the information on the heat of vaporization for water is irrelevant to this problem.)
$m L = m C \Delta T$
x (80 cal/g) = 100g 1cal/g°C 18°C
x (80 cal/g) = 1800 cal
x = 22.5 g

Only answer (D) matches this result.

37. **Heat transfer by electromagnetic waves is termed:**
     (Easy) (Skill 16.3)

     A. Conduction

     B. Convection

     C. Radiation

     D. Phase Change

Answer: C

To answer this question, recall the different ways that heat is transferred. Conduction is the transfer of heat through direct physical contact and molecules moving and hitting each other. Convection is the transfer of heat via density differences and flow of fluids. Radiation is the transfer of heat via electromagnetic waves (and can occur in a vacuum). Phase Change causes transfer of heat (though not of temperature) in order for the molecules to take their new phase. This is consistent, therefore, only with answer (C).

38. **A cooking thermometer in an oven works because the metals it is composed of have different:**
     (Average Rigor) (Skill 16.4)

     A. Melting points

     B. Heat convection

     C. Magnetic fields

     D. Coefficients of expansion

Answer: D

A thermometer of the type that can withstand oven temperatures works by having more than one metal strip. These strips expand at different rates with temperature increases, causing the dial to register the new temperature. This is consistent only with answer (D). If you did not know how an oven thermometer works, you could still omit the incorrect answers: It is unlikely that the metals in a thermometer would melt in the oven to display the temperature; the magnetic fields would not be useful information in this context; heat convection applies in fluids, not solids.

39. **Which of the following is not an assumption upon which the kinetic-molecular theory of gases is based?**
(Rigorous) (Skill 17.1)

   A. Quantum mechanical effects may be neglected

   B. The particles of a gas may be treated statistically

   C. The particles of the gas are treated as very small masses

   D. Collisions between gas particles and container walls are inelastic

Answer: D

Since the kinetic-molecular theory is classical in nature, quantum mechanical effects are indeed ignored, and answer A is incorrect. The theory also treats gases as a statistical collection of point-like particles with finite masses. As a result, answers B and C may also be eliminated. Thus, answer D is correct: collisions between gas particles and container walls are treated as elastic in the kinetic-molecular theory.

40. **What is temperature?**
(Average Rigor) (Skill 17.2)

   A. Temperature is a measure of the conductivity of the atoms or molecules in a material

   B. Temperature is a measure of the kinetic energy of the atoms or molecules in a material

   C. Temperature is a measure of the relativistic mass of the atoms or molecules in a material

   D. Temperature is a measure of the angular momentum of electrons in a material

Answer: B

Temperature is, in fact, a measure of the kinetic energy of the constituent components of a material. Thus, as a material is heated, the atoms or molecules that compose it acquire greater energy of motion. This increased motion results in the breaking of chemical bonds and in an increase in disorder, thus leading to melting or vaporizing of the material at sufficiently high temperatures.

41.   **Solids expand when heated because:**
       (Rigorous) (Skill 17.2)

       A. Molecular motion causes expansion

       B. PV = nRT

       C. Magnetic forces stretch the chemical bonds

       D. All material is effectively fluid

Answer: A

When any material is heated, the heat energy becomes energy of motion for the material's molecules. This increased motion causes the material to expand (or sometimes to change phase). Therefore, the answer is (A). Answer (B) is the ideal gas law, which gives a relationship between temperature, pressure, and volume for gases. Answer (C) is a red herring (misleading answer that is untrue). Answer (D) may or may not be true, but it is not the best answer to this question.

42. **What should be the behavior of an electroscope, which has been grounded in the presence of a positively charged object (1), after the ground connection is removed and then the charged object is removed from the vicinity (2)?**
(Average Rigor) (Skill 18.1)

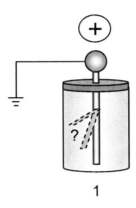

1   2

A. The metal leaf will start deflected (1) and then relax to an undeflected position (2)

B. The metal leaf will start in an undeflected position (1) and then be deflected (2)

C. The metal leaf will remain undeflected in both cases

D. The metal leaf will be deflected in both cases

Answer: B

When grounded, the electroscope will show no deflection. Nevertheless, if the ground is then removed and the charged object taken from the vicinity (in that order), the excess charge that existed near the sphere of the electroscope will distribute itself throughout the instrument, resulting in an overall net excess charge that will deflect the metal leaf.

43. **The electric force in Newtons, on two small objects (each charged to −10 microCoulombs and separated by 2 meters) is:**
    (Rigorous) (Skill 18.2)

    A. 1.0

    B. 9.81

    C. 31.0

    D. 0.225

Answer: D

To answer this question, use Coulomb's Law, which gives the electric force between two charged particles:
$F = k\, Q_1 Q_2 / r^2$
Then our unknown is F, and our knowns are:
$k = 9.0 \times 10^9\ Nm^2/C^2$
$Q_1 = Q_2 = -10 \times 10^{-6}\ C$
$r = 2\ m$

Therefore
$F = (9.0 \times 10^9)(-10 \times 10^{-6})(-10 \times 10^{-6})/(2^2)$ N
$F = 0.225$ N

This is compatible only with answer (D).

44.   **A 10 ohm resistor and a 50 ohm resistor are connected in parallel. If the current in the 10 ohm resistor is 5 amperes, the current (in amperes) running through the 50 ohm resistor is:**
        (Rigorous) (Skill 19.1)

        A. 1

        B. 50

        C. 25

        D. 60

Answer: A

To answer this question, use Ohm's Law, which relates voltage to current and resistance:
$V = IR$
where V is voltage; I is current; R is resistance.

We also use the fact that in a parallel circuit, the voltage is the same across the branches.

Because we are given that in one branch, the current is 5 amperes and the resistance is 10 ohms, we deduce that the voltage in this circuit is their product, 50 volts (from $V = IR$).

We then use $V = IR$ again, this time to find I in the second branch. Because V is 50 volts, and R is 50 ohm, we calculate that I has to be 1 ampere.

This is consistent only with answer (A).

45. **How much power is dissipated through the following resistive circuit?**
(Average Rigor) (Skill 19.4)

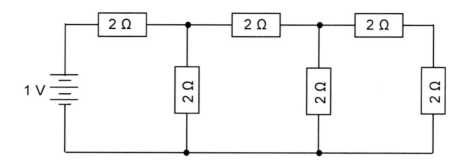

A. 0 W

B. 0.22 W

C. 0.31 W

D. 0.49 W

Answer: C

Use the rules of series and parallel resistors to quickly form an equivalent circuit with a single voltage source and a single resistor. In this case, the equivalent resistance is 3.25 Ω. The power dissipated by the circuit is the square of the voltage divided by the resistance. The final answer is C.

46. **The greatest number of 100 watt lamps that can be connected in parallel with a 120 volt system without blowing a 5 amp fuse is: (Rigorous) (Skill 19.4)**

    A. 24

    B. 12

    C. 6

    D. 1

Answer: C

To solve fuse problems, you must add together all the drawn current in the parallel branches, and make sure that it is less than the fuse's amp measure. Because we know that electrical power is equal to the product of current and voltage, we can deduce that:

$I = P/V$ (I = current (amperes); P = power (watts); V = voltage (volts))

Therefore, for each lamp, the current is 100/120 amperes, or 5/6 ampere. The highest possible number of lamps is thus six, because six lamps at 5/6 ampere each adds to 5 amperes; more will blow the fuse.

This is consistent only with answer (C).

47. **Which of the following statements may be taken as a legitimate inference based upon the Maxwell equation that states $\nabla \cdot \mathbf{B} = 0$? (Average Rigor) (Skill 20.1)**

    A. The electric and magnetic fields are decoupled

    B. The electric and magnetic fields are mediated by the W boson

    C. There are no photons

    D. There are no magnetic monopoles

Answer: D

Since the divergence of the magnetic flux density is always zero, there cannot be any magnetic monopoles (charges), given this Maxwell equation. If Gauss's law is applied to magnetic flux in the same manner as it is to electric flux, then the total magnetic "charge" contained within any closed surface must always be zero. This is another way of viewing the problem. Thus, answer D is correct. This answer may also be chosen by elimination of the other statements, which are untenable.

48. **What effect might an applied external magnetic field have on the magnetic domains of a ferromagnetic material?**
(Rigorous) (Skill 20.1)

    A. The domains that are not aligned with the external field increase in size, but those that are aligned decrease in size

    B. The domains that are not aligned with the external field decrease in size, but those that are aligned increase in size

    C. The domains align perpendicular to the external field

    D. There is no effect on the magnetic domains

Answer: B

Recall that ferromagnetic domains are portions of a magnetic material that have a local magnetic moment. The material may have an overall lack of a magnetic moment due to random alignment of its domains. In the presence of an applied field, the domains may align with the field to some extent, or the boundaries of the domains may shift to give greater weight to those domains that are aligned with the field, at the expense of those domains that are not aligned with the field. As a result, of the possibilities above, B is the best answer.

49. **What is the effect of running current in the same direction along two parallel wires, as shown below?**
(Rigorous) (Skill 20.4)

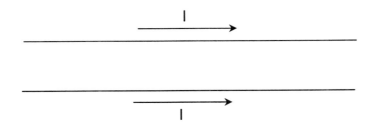

A. There is no effect

B. The wires attract one another

C. The wires repel one another

D. A torque is applied to both wires

Answer: B

Since the direction of the force on a current element is proportional to the cross product of the direction of the current element and the magnetic field, there is either an attractive or repulsive force between the two wires shown above. Using the right hand rule, it can be found that the magnetic field on the top wire due to the bottom wire is directed out of the plane of the page. Performing the cross product shows that the force on the upper wire is directed toward the lower wire. A similar argument can be used for the lower wire. Thus, the correct answer is B: an attractive force is exerted on the wires.

50.   **The current induced in a coil is defined by which of the following laws?**
        (Easy) (Skill 21.1)

        A. Lenz's Law

        B. Burke's Law

        C. The Law of Spontaneous Combustion

        D. Snell's Law

Answer: A

Lenz's Law states that an induced electromagnetic force always gives rise to a current whose magnetic field opposes the original flux change. There is no relevant "Snell's Law," "Burke's Law," or "Law of Spontaneous Combustion" in electromagnetism. (In fact, only Snell's Law is a real law of these three, and it refers to refracted light.) Therefore, the only appropriate answer is (A).

51.   **A light bulb is connected in series with a rotating coil within a magnetic field. The brightness of the light may be increased by any of the following except:**
        (Average Rigor) (Skill 21.1)

        A. Rotating the coil more rapidly.

        B. Using more loops in the coil.

        C. Using a different color wire for the coil.

        D. Using a stronger magnetic field.

Answer: C

To answer this question, recall that the rotating coil in a magnetic field generates electric current, by Faraday's Law. Faraday's Law states that the amount of emf generated is proportional to the rate of change of magnetic flux through the loop. This increases if the coil is rotated more rapidly (A), if there are more loops (B), or if the magnetic field is stronger (D). Thus, the only answer to this question is (C).

52. **What is the direction of the magnetic field at the center of the loop of current (I) shown below (i.e., at point A)?**
(Easy) (Skill 21.2)

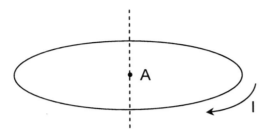

A. Down, along the axis (dotted line)

B. Up, along the axis (dotted line)

C. The magnetic field is oriented in a radial direction

D. There is no magnetic field at point A

Answer: A

The magnetic field may be found by applying the right-hand rule. The magnetic field curls around the wire in the direction of the curled fingers when the thumb is pointed in the direction of the current. Since there is a degree of symmetry, with point A lying in the center of the loop, the contributions of all the current elements on the loop must yield a field that is either directed up or down at the axis. Use of the right-hand rule indicates that the field is directed down. Thus, answer A is correct.

53. **The use of two circuits next to each other, with a change in current in the primary circuit, demonstrates:**
(Rigorous) (Skill 21.4)

    A. Mutual current induction

    B. Dielectric constancy

    C. Harmonic resonance

    D. Resistance variation

Answer: A

To answer this question, recall that changing current induces a change in magnetic flux, which in turn causes a change in current to oppose that change (Lenz's and Faraday's Laws). Thus, (A) is correct. If you did not remember that, note that harmonic resonance is irrelevant here (eliminating (C)), and there is no change in resistance in the circuits (eliminating (D)).

54. **A semi-conductor allows current to flow:**
(Easy) (Skill 22.1)

    A. Never

    B. Always

    C. As long as it stays below a maximum temperature

    D. When a minimum voltage is applied

Answer: D

To answer this question, recall that semiconductors do not conduct as well as conductors (eliminating answer (B)), but they conduct better than insulators (eliminating answer (A)). Semiconductors can conduct better when the temperature is higher (eliminating answer (C)), and their electrons move most readily under a potential difference. Thus the answer can only be (D).

55. **All of the following use semi-conductor technology, except a(n):**
    (Average Rigor) (Skill 22.1)

    A. Transistor

    B. Diode

    C. Capacitor

    D. Operational Amplifier

Answer: C

Semi-conductor technology is used in transistors and operational amplifiers, and diodes are the basic unit of semi-conductors. Therefore the only possible answer is (C), and indeed a capacitor does not require semi-conductor technology.

56. **A wave generator is used to create a succession of waves. The rate of wave generation is one every 0.33 seconds. The period of these waves is:**
    (Average Rigor) (Skill 23.1)

    A. 2.0 seconds

    B. 1.0 seconds

    C. 0.33 seconds

    D. 3.0 seconds

Answer: C

The definition of a period is the length of time between wave crests. Therefore, when waves are generated one per 0.33 seconds, that same time (0.33 seconds) is the period. This is consistent only with answer (C). Do not be trapped into calculating the number of waves per second, which might lead you to choose answer (D).

57. **An electromagnetic wave propagates through a vacuum. Independent of its wavelength, it will move with constant:** *(Easy) (Skill 23.2)*

    A. Acceleration

    B. Velocity

    C. Induction

    D. Sound

Answer: B

Electromagnetic waves are considered always to travel at the speed of light, so answer (B) is correct. Answers (C) and (D) can be eliminated in any case, because induction is not relevant here, and sound does not travel in a vacuum.

58. **A wave has speed 60 m/s and wavelength 30,000 m. What is the frequency of the wave?** *(Average Rigor) (Skill 23.2)*

    A. 2.0×10^{-3} Hz

    B. 60 Hz

    C. 5.0×10^{2} Hz

    D. 1.8×10^{6} Hz

Answer: A

To answer this question, recall that wave speed is equal to the product of wavelength and frequency. Thus:
60 m/s = (30,000 m) (frequency)
frequency = 2.0×10^{-3} Hz

This is consistent only with answer (A).

59.　**Rainbows are created by:**
　　　(Easy) (Skill 24.1)

　　　A.　Reflection, dispersion, and recombination

　　　B.　Reflection, resistance, and expansion

　　　C.　Reflection, compression, and specific heat

　　　D.　Reflection, refraction, and dispersion

Answer: D

To answer this question, recall that rainbows are formed by light that goes through water droplets and is dispersed into its colors. This is consistent with both answers (A) and (D). Then note that refraction is important in bending the differently colored light waves, while recombination is not a relevant concept here. Therefore, the answer is (D).

60.　**Which of the following is *not* a legitimate explanation for refraction of light rays at boundaries between different media?**
　　　(Rigorous) (Skill 24.1)

　　　A.　Light seeks the path of least time between two different points

　　　B.　Due to phase matching and other boundary conditions, plane waves travel in different directions on either side of the boundary, depending on the material parameters

　　　C.　The electric and magnetic fields become decoupled at the boundary

　　　D.　Light rays obey Snell's law

Answer: C

Even if the exact implications of each explanation are not known or understood, answer C can be chosen due to its plain incorrectness. The other responses involve more or less fundamental explanations for the refraction of light rays (which are equivalent to plane waves) at media boundaries.

61. A stationary sound source produces a wave of frequency *F*. An observer at position A is moving toward the horn, while an observer at position B is moving away from the horn. Which of the following is true?
(*Rigorous*) (*Skill 24.2*)

    A. $F_A < F < F_B$

    B. $F_B < F < F_A$

    C. $F < F_A < F_B$

    D. $F_B < F_A < F$

Answer: B

To answer this question, recall the Doppler Effect. As a moving observer approaches a sound source, s/he intercepts wave fronts sooner than if s/he were standing still. Therefore, the wave fronts seem to be coming more frequently. Similarly, as an observer moves away from a sound source, the wave fronts take longer to reach him/her. Therefore, the wave fronts seem to be coming less frequently. Because of this effect, the frequency at B will seem lower than the original frequency, and the frequency at A will seem higher than the original frequency. The only answer consistent with this is (B). Note also, that even if you weren't sure of which frequency should be greater/smaller, you could still reason that A and B should have opposite effects, and be able to eliminate answer choices (C) and (D).

62.   A monochromatic ray of light passes from air to a thick slab of glass
        (n = 1.41) at an angle of 45° from the normal. At what angle does it
        leave the air/glass interface?
        (Rigorous) (Skill 24.3)

            A. 45°

            B. 30°

            C. 15°

            D. 55°

Answer: B

To solve this problem use Snell's Law:
$n_1 \sin\theta_1 = n_2 \sin\theta_2$ (where n_1 and n_2 are the indexes of refraction and θ_1
and θ_2 are the angles of incidence and refraction).

Then, since the index of refraction for air is 1.0, we deduce:
$1 \sin 45° = 1.41 \sin x$
$x = \sin^{-1} ((1/1.41) \sin 45°)$
$x = 30°$

This is consistent only with answer (B). Also, note that you could
eliminate answers (A) and (D) in any case, because the refracted light will
have to bend at a smaller angle when entering glass.

63. **If one sound is ten decibels louder than another, the ratio of the intensity of the first to the second is:**
(Average Rigor) (Skill 25.1)

    A. 20:1

    B. 10:1

    C. 1:1

    D. 1:10

Answer: B

To answer this question, recall that a decibel is defined as ten times the log of the ratio of sound intensities:
(decibel measure) = $10 \log (I / I_0)$ where I_0 is a reference intensity.

Therefore, in our case,
(decibels of first sound) = (decibels of second sound) + 10
$10 \log (I_1 / I_0) = 10 \log (I_2 / I_0) + 10$
$10 \log I_1 - 10 \log I_0 = 10 \log I_2 - 10 \log I_0 + 10$
$10 \log I_1 - 10 \log I_2 = 10$
$\log (I_1 / I_2) = 1$
$I_1 / I_2 = 10$

This is consistent only with answer (B).
(Be careful not to get the two intensities confused with each other.)

64.   The velocity of sound is greatest in:
       (Average Rigor) (Skill 25.2)

           A. Water

           B. Steel

           C. Alcohol

           D. Air

Answer: B

Sound is a longitudinal wave, which means that it shakes its medium in a way that propagates as sound traveling. The speed of sound depends on both elastic modulus and density, but for a comparison of the above choices, the answer is always that sound travels faster through a solid like steel, than through liquids or gases. Thus, the answer is (B).

65.   A vibrating string's frequency is _____ proportional to the

       _____.
       (Rigorous) (Skill 25.2)

           A. Directly; Square root of the tension

           B. Inversely; Length of the string

           C. Inversely; Squared length of the string

           D. Inversely; Force of the plectrum

Answer: A

To answer this question, recall that
$f = (n\,v) / (2\,L)$ where f is frequency; v is velocity; L is length

and

$v = (F_{tension} / (m / L))^{\frac{1}{2}}$ where $F_{tension}$ is tension; m is mass; others as above

so

$f = (n / 2\,L)\,((F_{tension} / (m / L))^{\frac{1}{2}})$

indicating that frequency is directly proportional to the square root of the tension force. This is consistent only with answer (A). Note that in the final frequency equation, there is an inverse relationship with the square root of the length (after canceling like terms). This is not one of the options, however.

66. **Which of the following apparatus can be used to measure the wavelength of a sound produced by a tuning fork?**
(Average Rigor) (Skill 25.3)

    A. A glass cylinder, some water, and iron filings

    B. A glass cylinder, a meter stick, and some water

    C. A metronome and some ice water

    D. A comb and some tissue

Answer: B

To answer this question, recall that a sound will be amplified if it is reflected back to cause positive interference. This is the principle behind musical instruments that use vibrating columns of air to amplify sound (e.g. a pipe organ). Therefore, presumably a person could put varying amounts of water in the cylinder, and hold the vibrating tuning fork above the cylinder in each case. If the tuning fork sound is amplified when put at the top of the column, then the length of the air space would be an integral multiple of the sound's wavelength. This experiment is consistent with answer (B). Although the experiment would be tedious, none of the other options for materials suggest a better alternative.

67. **The highest energy is associated with:**
(Easy) (Skill 26.1)

    A. UV radiation

    B. Yellow light

    C. Infrared radiation

    D. Gamma radiation

Answer: D

To answer this question, recall the electromagnetic spectrum. The highest energy (and therefore frequency) rays are those with the lowest wavelength, i.e. gamma rays. (In order of frequency from lowest to highest are: radio, microwave, infrared, red through violet visible light, ultraviolet, X-rays, gamma rays.) Thus, the only possible answer is (D). Note that even if you did not remember the spectrum, you could deduce that gamma radiation is considered dangerous and thus might have the highest energy.

68.   **An object two meters tall is speeding toward a plane mirror at 10 m/s. What happens to the image as it nears the surface of the mirror?**
      (Rigorous) (Skill 27.1)

      A. It becomes inverted.

      B. The Doppler Effect must be considered.

      C. It remains two meters tall.

      D. It changes from a real image to a virtual image.

Answer: C

Note that the mirror is a plane mirror, so the image is always a virtual image of the same size as the object. If the mirror were concave, then the image would be inverted until the object came within the focal distance of the mirror. The Doppler Effect is not relevant here. Thus, the only possible answer is (C).

69.   **Automobile mirrors that have a sign, "objects are closer than they appear" say so because:**
      (Rigorous) (Skill 27.2)

      A. The real image of an obstacle, through a converging lens, appears farther away than the object.

      B. The real or virtual image of an obstacle, through a converging mirror, appears farther away than the object.

      C. The real image of an obstacle, through a diverging lens, appears farther away than the object.

      D. The virtual image of an obstacle, through a diverging mirror, appears farther away than the object.

Answer: D

To answer this question, first eliminate answer choices (A) and (C), because we have a mirror, not a lens. Then draw ray diagrams for diverging (convex) and converging (concave) mirrors, and note that because the focal point of a diverging mirror is behind the surface, the image is smaller than the object. This creates the illusion that the object is farther away, and therefore (D) is the correct answer.

70.   If an object is 20 cm from a convex lens whose focal length is 10 cm, the image is:
(Rigorous) (Skill 27.2)

A. Virtual and upright

B. Real and inverted

C. Larger than the object

D. Smaller than the object

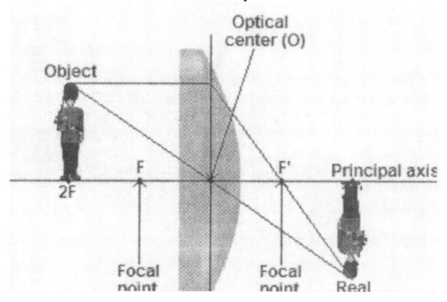

Answer: B

To solve this problem, draw a lens diagram with the lens, focal length, and image size.

The ray from the top of the object straight to the lens is focused through the far focus point; the ray from the top of the object through the near focus goes straight through the lens; the ray from the top of the object through the center of the lens continues. These three meet to form the "top" of the image, which is therefore real and inverted. This is consistent only with answer (B).

71. The constant of proportionality between the energy and the frequency of electromagnetic radiation is known as the: *(Easy) (Skill 28.1)*

>   A. Rydberg constant
>
>   B. Energy constant
>
>   C. Planck constant
>
>   D. Einstein constant

Answer: C

Planck estimated his constant to determine the ratio between energy and frequency of radiation. The Rydberg constant is used to find the wavelengths of the visible lines on the hydrogen spectrum.
The other options are not relevant options, and may not actually have physical meaning. Therefore, the only possible answer is (C).

72. Which phenomenon was first explained using the concept of quantization of energy, thus providing one of the key foundational principles for the later development of quantum theory? *(Rigorous) (Skill 28.1)*

>   A. The photoelectric effect
>
>   B. Time dilation
>
>   C. Blackbody radiation
>
>   D. Magnetism

Answer: C

Although the photoelectric effect applied principles of quantization in explaining the behavior of electrons emitted from a metallic surface when the surface is illuminated with electromagnetic radiation, the explanation of the phenomenon of blackbody radiation, provided by Max Planck, was the first major success of the concept of quantized energy. Magnetism may be explained quantum mechanically, but such an explanation was not forthcoming until well after Planck's quantization hypothesis. Time dilation is primarily explained through relativity theory.

73. **Which statement best describes why population inversion is necessary for a laser to operate?**
(Rigorous) (Skill 28.3)

   A. Population inversion prevents too many electrons from being excited into higher energy levels, thus preventing damage to the gain medium.

   B. Population inversion maintains a sufficient number of electrons in a higher energy state so as to allow a significant amount of stimulated emission.

   C. Population inversion prevents the laser from producing coherent light.

   D. Population inversion is not necessary for the operation of most lasers.

Answer: B

Population inversion is a state in which there are a larger number of electrons in a particular higher-energy excited state than in a particular lower-energy state. When perturbed by a passing photon, these electrons may then emit a photon of the same energy (frequency) and phase. This is the process of stimulated emission, which, when population inversion is obtained, can produce something of a "chain reaction," thus giving lasers their characteristically monochromatic and highly coherent light.

74. **Bohr's theory of the atom was the first to quantize:**
(Average Rigor) (Skill 29.1)

   A. Work

   B. Angular Momentum

   C. Torque

   D. Duality

Answer: B

Bohr was the first to quantize the angular momentum of electrons, as he combined Rutherford's planet-style model with his knowledge of emerging quantum theory. Recall that he derived a "quantum condition" for the single electron, requiring electrons to exist at specific energy levels

75. **Two neutral isotopes of a chemical element have the same numbers of:**
(Easy) (Skill 29.2)

    A. Electrons and Neutrons

    B. Electrons and Protons

    C. Protons and Neutrons

    D. Electrons, Neutrons, and Protons

Answer: B

To answer this question, recall that isotopes vary in their number of neutrons. (This fact alone eliminates answers (A), (C), and (D).) If you did not recall that fact, note that we are given that the two samples are of the same element, constraining the number of protons to be the same in each case. Then, use the fact that the samples are neutral, so the number of electrons must exactly balance the number of protons in each case. The only correct answer is thus (B).

76. **When a radioactive material emits an alpha particle only, its atomic number will:**
(Average Rigor) (Skill 30.1)

    A. Decrease

    B. Increase

    C. Remain unchanged

    D. Change randomly

Answer: A

To answer this question, recall that in alpha decay, a nucleus emits the equivalent of a Helium atom. This includes two protons, so the original material changes its atomic number by a decrease of two.

77. Ten grams of a sample of a radioactive material (half-life = 12 days) were stored for 48 days and re-weighed. The new mass of material was:
(Rigorous) (Skill 30.2)

    A. 1.25 g

    B. 2.5 g

    C. 0.83 g

    D. 0.625 g

Answer: D

To answer this question, note that 48 days is four half-lives for the material. Thus, the sample will degrade by half four times. At first, there are ten grams, then (after the first half-life) 5 g, then 2.5 g, then 1.25 g, and after the fourth half-life, there remains 0.625 g. You could also do the problem mathematically, by multiplying ten times $(½)^4$, i.e. ½ for each half-life elapsed.

78. Which of the following pairs of elements are not found to fuse in the centers of stars?
(Average Rigor) (Skill 31.1)

    A. Oxygen and Helium

    B. Carbon and Hydrogen

    C. Beryllium and Helium

    D. Cobalt and Hydrogen

Answer: D

To answer this question, recall that fusion is possible only when the final product has more binding energy than the reactants. Because binding energy peaks near a mass number of around 56, corresponding to Iron, any heavier elements would be unlikely to fuse in a typical star. (In very massive stars, there may be enough energy to fuse heavier elements.) Of all the listed elements, only Cobalt is heavier than iron, so answer (D) is correct.

79.   **In a fission reactor, heavy water:**
       (Average Rigor) (Skill 31.2)

       A. Cools off neutrons to control temperature

       B. Moderates fission reactions

       C. Initiates the reaction chain

       D. Dissolves control rods

Answer: B

In a nuclear reactor, heavy water is made up of oxygen atoms with hydrogen atoms called 'deuterium,' which contain two neutrons each. This allows the water to slow down (moderate) the neutrons, without absorbing many of them. This is consistent only with answer (B).

80. **Given the following values for the masses of a proton, a neutron and an alpha particle, what is the nuclear binding energy of an alpha particle?**
(*Rigorous*) *(Skill 31.3)*

Proton mass = 1.6726×10^{-27} kg
Neutron mass = 1.6749×10^{-27} kg
Alpha particle mass = 6.6465×10^{-27} kg

    A. 0 J

    B. 7.3417×10^{-27} J

    C. 4 J

    D. 4.3589×10^{-12} J

Answer: D

The nuclear binding energy is the amount of energy that is required to break the nucleus into its component nucleons. In this case, the binding energy of an alpha particle, which is composed of two protons and two neutrons, is calculated by first finding the difference between the sum of the masses of all the nucleons and the mass of the alpha particle. Using the equation $E = mc^2$ to find the energy in terms of the mass difference of 4.85×10^{-29} kg, and using the speed of light of about 2.9979×10^8 m/s, the result is the value given in answer D.

Free Response Questions: Sample Responses

Question 1:

A 2 meter square loop of wire, connected to a resistive load, is arranged near a power line carrying 100 Amperes of current at 60 Hertz, as shown below. Assuming that the load is negligible in size, calculate the voltage amplitude across the load.

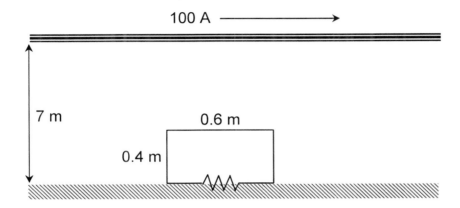

Answer:

Varying magnetic fields produce an electromotive force (EMF) in a closed loop of wire (a circuit). This EMF, in turn, can supply current to a load in the circuit. In the diagram above, energy is linked from the power line (a current-carrying conductor) to the wire loop through the time-varying magnetic field. The problem, as stated, involves calculating the magnetic flux through the loop and then the resulting EMF across the load.

First, it is necessary to calculate the magnetic field as a function of distance from the wire. It is safe to assume that the power line is long, and therefore it may be treated approximately as an infinitely long line of current. Using the cylindrical symmetry of the wire and applying Ampere's circuital law, the magnetic flux density B may be calculated as a function of distance from the wire. It is assumed that the flux density does not vary in the horizontal direction (along the length of the wire).

$$\oint_C \mathbf{B} \cdot d\mathbf{l} = \mu_0 I$$

where I is the current through the wire and $\mu_0 = 4\pi \times 10^{-7} \, N / A^2$ is the permeability of free space.

PHYSICS

Choosing a circular path C of radius r centered on the power line, and noting that the flux density is invariant along the path, yields:

$$2\pi r B = \mu_0 I$$

$$B = \frac{\mu_0 I}{2\pi r}$$

By the right-hand rule, the flux is directed into the page at all points in the plane of the wire loop. Next, calculate the total flux Ψ through the loop:

$$\Psi = \iint_S B \, dS = 0.6m \cdot \int_{7-0.4}^{7.} \frac{\mu_0 I}{2\pi r} dr$$

$$\Psi = \iint_S B \, dS = 0.6m \frac{\mu_0 I}{2\pi}[\ln(7) - \ln(6.6)]$$

Calculating the numerical value yields:

$$\Psi = 7.06 \times 10^{-7} \, Wb$$

Note that the EMF induced is proportional to the derivative of the flux with respect to time. This may be derived from Maxwell's equations using Stokes' theorem.

$$EMF = -\frac{d}{dt}\Psi$$

Since power lines involve sinusoidally varying currents, the induced EMF should simply be the flux through the wire multiplied by the angular frequency, $\omega = 2\pi f$.

$$EMF = 2\pi(60Hz)(7.06 \times 10^{-7} \, Wb) = \boxed{2.66 \times 10^{-4} \, V}$$

This is the voltage (EMF) induced in the wire due to the presence of the power line, and, therefore, it is also the voltage across the load. Since the direction of induced current flow is not relevant to this problem, the absolute value of the EMF is given as the solution.

Question 2:

A 2,000 kg space vessel is designed to be accelerated away from the Sun using radiation pressure exerted on a sail of area 100 m^2 that has 95% reflectivity. (Assume that the sail absorbs the rest of the photons.) If the photon flux is assumed to be uniform over the space of interest, and if it has nominal average values of about 3.8x10^{21} photons m^{-2} s^{-1} and 5.5x10^{14} Hz frequency, what is the acceleration of the space vessel?

Answer:

Radiation pressure is the force exerted on an object due to the absorption or reflection of photons. Typically, this force is not noticeable, since it is generally very small. The lack of air resistance in space, combined with the large photon flux produced by the Sun, makes radiation pressure a potential source of propulsion for space travel.

To calculate the acceleration of the space vessel, first notice that the photons have momentum p. This momentum, according to the following expression, can be calculated based on the wave properties of light.

$$p = \frac{h}{\lambda}$$

Alternatively:

$$p = \frac{E}{c} = \frac{h\nu}{c}$$

The force on the space vessel is due to the absorption or reflection of photons. The momentum of the massless photons is imparted to the vessel (through the sail) when absorption takes place, and twice the momentum is imparted to the vessel when reflection takes place. Thus, 95% of the flux incident on the sail will impart twice its associated momentum, and 5% of the flux will impart simply the single value of its momentum. The average total momentum imparted to the sail, per second, is then:

$$\Delta p = \left[0.95 \times 2\frac{h\nu_{average}}{c} + 0.05\frac{h\nu_{average}}{c} \right]\frac{1}{photon}\left(3.8 \times 10^{21}\frac{photons}{m^2} \cdot 100\,m^2 \right)$$

Here, the photon flux is multiplied by the distribution of momentum for reflected and absorbed photons and by the area of the sail. Using the values for h, c and $v_{average}$ yields the following result:

$$\Delta p = 7.41 \times 10^{23} \frac{\left(6.626 \times 10^{-34} \, J \cdot s\right)\left(5.5 \times 10^{14} \, Hz\right)}{3.00 \times 10^8 \, \frac{m}{s}} = 9.00 \times 10^{-4} \frac{kg \cdot m}{s}$$

The force is the time rate of change of the momentum:

$$F = \frac{\Delta p}{\Delta t}$$

The product of the acceleration and the mass of the space vessel may then be equated to the force. This yields the final answer when the numerical calculation is performed.

$$F = ma = \frac{\Delta p}{\Delta t}$$

$$a = \frac{\Delta p}{m \Delta t}$$

Since Δt is simply 1 second here, and the mass of the space vessel is 2,000 kg, the final answer is:

$$a = \frac{\Delta p}{m \cdot 1s}$$

$$a = \frac{9.00 \times 10^{-4} \frac{kg \cdot m}{s}}{(2,000 \, kg)(1 \, s)} = \boxed{4.50 \times 10^{-7} \frac{m}{s^2}}$$

This is an extremely small value, which implies that either the sail must be made much larger, or the design must be abandoned due to a lack of realistic propulsion power from radiation pressure.

MASSACHUSETTS TEST FOR EDUCATOR LICENTURE - MTEL - 2008

PO#   Store/School:

Address 1:

Address 2 (Ship to other):

City, State Zip

Credit card number_____-_____-_____-_____   expiration_____

EMAIL _____

PHONE   **FAX**

ISBN	TITLE	Qty	Retail	Total
978-1-58197-287-0	MTEL Communication and Literacy Skills 01			
978-1-58197-592-5	MTEL General Curriculum (formerly Elementary) 03			
978-1-58197-607-6	MTEL History 06 (Social Science)			
978-1-58197-283-2	MTEL English 07			
978-1-58197-349-5	MTEL Mathematics 09			
978-1-58197-593-2	MTEL General Science 10			
978-1-58197-041-8	MTEL Physics 11			
978-1-58197-883-4	MTEL Chemistry 12			
978-1-58197-687-8	MTEL Biology 13			
978-1-58197-683-0	MTEL Earth Science 14			
978-1-58197-676-2	MTEL Early Childhood 02			
978-1-58197-893-3	MTEL Visual Art Sample Test 17			
978-1-58197-8988	MTEL Political Science/ Political Philosophy 48			
978-1-58197-886-5	MTEL Physical Education 22			
978-1-58197-887-2	MTEL French Sample Test 26			
978-1-58197-888-9	MTEL Spanish 28			
978-1-58197-889-6	MTEL Middle School Mathematics 47			
978-1-58197-890-2	MTEL Middle School Humanities 50			
978-1-58197-891-9	MTEL Middle School Mathematics-Science 51			
978-1-58197-266-5	MTEL Foundations of Reading 90 (requirement all El. Ed)			
			SUBTOTAL	
			Ship	$8.25
			TOTAL	

PHYSICS   286